The Future of Amazonia in Brazil

PETER LANG
PROMPT

PETER LANG
New York • Bern • Berlin
Brussels • Vienna • Oxford • Warsaw

Marcílio de Freitas and
Marilene Corrêa da Silva Freitas

The Future of Amazonia in Brazil

A Worldwide Tragedy

PETER LANG
New York • Bern • Berlin
Brussels • Vienna • Oxford • Warsaw

Library of Congress Control Number: 2019058475

Bibliographic information published by **Die Deutsche Nationalbibliothek.**
Die Deutsche Nationalbibliothek lists this publication in the "Deutsche Nationalbibliografie"; detailed bibliographic data are available on the Internet at http://dnb.d-nb.de/.

ISBN 978-1-4331-7793-4 (hardcover)
ISBN 978-1-4331-7790-3 (ebook pdf)
ISBN 978-1-4331-7791-0 (epub)
ISBN 978-1-4331-7792-7 (mobi)
DOI 10.3726/b16717

This publication has been peer reviewed and meets the highest quality standards for content and production.

The paper in this book meets the guidelines for permanence and durability of the Committee on Production Guidelines for Book Longevity of the Council of Library Resources.

Printed in the United States of America

We dedicate this book to our brother friends Argemiro Gomes de Andrade Júnior and Maria de Nazaré Corrêa da Silva who sowed solidarity and joy during their lives (in memoriam).

Table of Contents

Preface

The objective of this book is to disseminate and popularize Amazonia's importance for people and mankind. It also presents denunciations on the destructive interventions made by Brazil's new president on the use and protection policies of Amazonia.

We hope that readers will increase their efforts in Amazonia's defense and protection, at all times, in all places. Its material and symbolic importance for Brazil and the world goes beyond the Brazilian state's political interests. Its geography, history, and cultural and theological significance are intertwined with universal history.

Through a prospective study, structural issues that affect Amazonia's ecological and cultural complexity and its globalization are analyzed in this book. Its insertion in world geopolitics permeated by sustainability is presented through macro-scenarios. The use of macro-scenarios in the description and analysis of the problems presented in this book open the way to a more consistent and systemic interdisciplinary approach. What is lost in detail in the analysis of the focal problems is gained in integration and interpretation. As a result, we have book's better thematic unity. This methodological strategy allows us to identify and apprehend the heuristic range of the book's proposal, and its main contributions more quickly. It is organized in nine chapters.

The first is a brief retrospective study on its political geography, environmental scenarios, and links with the world. Some of the scenarios predicted for the environment, Amazonia, and sustainability are reviewed, raising the question of science education in terms of "What is Amazonia's role in the world? And what is the world's role in Amazonia?" The second chapter presents the globalization of environmental issues and the uncertainties and controversies of the twenty-first century. The third is centered on the utopias of the twenty-first century. Contradictions in Western culture are analyzed. The scientific, economic, and political connections between sustainable development and the places, regions, and the world are shown. The relations between Amazonia and the new world contracts are presented, reaffirming again Amazonia's importance to mankind. The fourth chapter analyzes the nexus between science, religion and the origins of universe and life within the context of sustainability. It articulates science and religion emphasizing the contradictions, ruptures, and links with sustainability and Amazonia. Scenarios that discuss Amazonia's ecological, theological, and scientific importance to the current civilization framework are presented. It also presents the importance of sustainability to citizen education, mankind, and the planet's future. Finally, it concludes with important considerations on the insertion of sustainability in social, economic, scientific, and religious processes. The fifth chapter reaffirms sustainability as the main invention of the twenty-first century. Exploratory studies are presented on the aspects of processes of Western civilization, which have contributed to sustainability as fundamental to the solution of postmodernity's complex issues. It proposes to expand the significance of the concept of sustainability and its application to scenarios immersed in the nature-culture universe. The sixth chapter presents the predatory development of Amazonia dating from the beginning of the twentieth century. It shows that capitalism does not have the heuristic reach to economically exploit Amazonia while preserving its biomes. Proposals and uncertainties concerning Amazonia's ecological issues are analyzed. Environmental and social impacts from large socioeconomic development projects implemented in the region are shown. Technical elements to clarify the sustainability concept and its correlation with the development of Amazonia and the world are presented and analyzed. In the seventh chapter , several actions taken by Brazil's new president designed to destroy important sustainable programs from Amazonia and Brazil are presented. It shows politics in the service of non-sustainability. Environmental and social tragedies caused by large mining companies are also analyzed. The study reaffirms the incompatibility of predatory capitalist productive forces with the sustainability of places and people. It also shows the control that these productive forces have over the national state, turning it into a weapon against citizenship. Through macro-scenarios, the eighth chapter shows from a humanist perspective, that the history of capitalist and communist expansion was a tragedy. To this day, capitalism continues destroying the environments and cultures of colonized peoples. Its wars and the unsustainable use of the planet are a growing threat to mankind's future. In bloodthirsty actions, the use of weapons always surpasses diplomacy and sustainability. The Vietnam War remains as a singular example.

In a slower and camouflaged but no less predatory form, the repetition of this history is fully under way in Amazonia. The Brazilian state is its main protagonist. To illustrate this framework, several scientific and ethnological scenarios from Brazilian Amazonia are presented, and the importance of this region to contemporary world is emphasized. Finally, in the ninth chapter, our commitment to sustainability and the preservation of Amazonia are presented, along with the six main issues raised by the twenty-first century's concern with sustainable development. All related to the accumulation, expansion, and circulation of capital.

As proposed by the authors, this book has broadened the meaning of the notion of sustainability and reaffirmed the importance of Amazonia for Brazil and mankind. In this complex panel, it presents new theoretical and empirical elements in order to understand Amazonia and its links with sustainability, culture, and nature. It also constitutes a complaint against the destruction of Amazonia and an instrument for environmental preservation and education for sustainability. Brazilian Amazonia is in jeopardy. This potential tragedy has accelerated this publication.

Manaus, October, 14, 2019
Marcílio de Freitas

Foreword

Amazonia is one of the planet's last utopias. Before the New World was discovered, it already instigated the imaginary of people, travelers, and government. What is its future? *The Future of Amazonia in Brazil: A Worldwide Tragedy* is a study on the importance of protecting Amazonia and constructing its sustainable development.

According to Brazil's new president, Jair Bolsonaro, "Environmental preservation should not be important for those who want to do business and produce." At this juncture, at this time, Amazonia is a great border open to the predatory forces of capitalist production. Its continued existence is in danger. On the initiative of the UN, World Environment Day has been celebrated on 5 June since 1972. Its objective is to promote programs and actions that protect and preserve the environment, raise awareness on the need to protect life, the planet, and the future of mankind. The world environment has never been more threatened. This study approaches this subject of universal importance with Amazonia and worldwide processes as its central focus. It is organized on the basis of recent research and studies I have developed in collaboration with Marilene Corrêa da Silva Freitas since 2017. Its central axis is sustainable development in Amazonia and its articulations with worldwide socioeconomic processes. The authors integrate several themes that compose the theoretical and empirical substratum of sustainability, in the context of the development of Amazonia and worldwide processes, as illustrated in Figure 1. The destruction of Amazonia announced by the current Brazilian government has

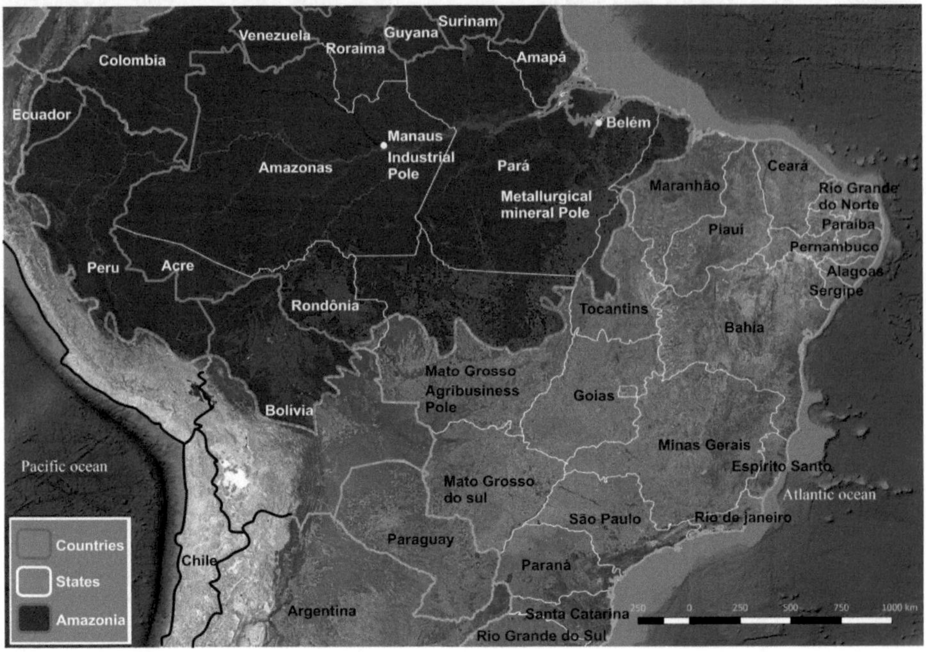

Figure 1. Map of Amazonia with its two main cities, Manaus and Belem, and its main economic poles.

accelerated the organization of this book. The threat to 68% of its protected areas and indigenous territories, this is equivalent to 390 million hectares (Barifouse, 2019), the state's breaking with international import agreements on the ecological protection of Amazonia, and the complete opening up of the region to predatory capitalist development are the Brazilian government's perverse political actions. It is a tragic picture, the return of the massive timber and mineral complexes, agroforestry companies, deforestation (cutting and burning) and the occupation for extensive agriculture and cattle breeding, the expansion of agribusiness, multiplication of huge hydroelectric plants, and infrastructure for highways, in addition to public policies on the research and exploration of fossil energy and natural gas, causing irreversible environmental destruction on Amazonia. Inadequate environmental protection from these initiatives will accelerate the irreversible ecological destruction of the region. These are the main threats to Amazonia, all of them legitimated by the current Brazilian government. In 2018, Brazil led the world in deforestation, with the disappearance of 1.3 million hectares of its primary forests (McGrath, 2019). In addition, the Report published by the UN registers the threat of extinction of 1 million species (UN News, 2019). Life has never been so threatened on the planet. The destructive interventions in Amazonia lead this civilizing barbarism that affects us all. Man's relationships with nature need to change radically. Simultaneously, there is need for a radical break with predatory capitalism. At its limit, it is paramount to give new meaning to the foundations, the explanatory meanings, and the operative mechanisms of this type of capitalism. We can affirm that, in Amazonia, the growing poverty of its populations is proportionate to the degree of destruction of its cultures and biomes. These subjects are analyzed in this book. New proposals are made with a view to safeguarding life and nonpredatory development; for a better and sustainable world for us all, at this time and in the future.

The second decade of this century has reaffirmed the importance of ecology to the processes underlying civilization. The environmental degradation of the planet is putting mankind's future at risk. The environmental destruction of Amazonia deepens this crisis, and generates worldwide socioeconomic impacts. Changing the polluting industrial matrices and the relationship between man and nature is a mark of this new era. In a certain way, the solution of complex problems such as the matrices of occupation, production, and social dynamics of places and planets presupposes a new design and socioeconomic insertion of people in the contemporary world. This context has formalized the need to construct sustainable processes.

In counterpoint to the global economy's pragmatic structures, worldwide public opinion builds new ethical foundations committed to the human condition and to universal rights. The overlapping of world issues with regional development in a process of mutual feedback has required a continuous recreation of social and economic realities, resonating in social groups, in general, and in the human psyche, in particular.

In spite of global environmental protection policies, deterioration of worldwide biodiversity has been aggravated at an increasing speed. Its main causes are the use of fossil fuel; inadequate use of soils and waters; commercial overexploitation of some species; introduction of predatory species in certain ecosystems; increasing soil, air, and water pollution; intensification of extensive agriculture using predatory techniques; territorial reordering and global climate change; and lack of political consensus among the world's major nations on the implementation of ecological protection measures for the planet. The United States of America and, more recently, Brazil lead this political resistance.

Various other factors are also contributing to this process, among which are the acceleration of demographic growth, policies of forcing nonadaptive and nonintegrated economic development on local and regional environments, nonregulation of the rights of access to the natural resources; and insufficiency of scientific knowledge on regional and worldwide ecological dynamics.

Promotion of sustainable development depends on the solution of complex problems, among which are the development and consumption of efficient non-polluting energy sources; reorganization of the land transport sector and better management of traffic systems; substitution of the industrial pollution matrix; protection of the sea's natural resources and of soil and air use; institutionalization of the mechanisms of measurement and control of air pollution; better management of the impacts of climate change; combating noise pollution; better management and protection of water resources; preservation and adjusted management of biodiversity and natural heritage; development of mechanisms to minimize risks and protect human health in unhealthy occupational matrices; better management and control of ecotoxicology and the impact of fungicides and pesticides; mobilization of the theoretical and empirical structures of the economic and social sciences; development of human resources for the management of sustainable development; and elimination of extreme poverty.

The transformations of these dynamic variables into quantitative indicators to be incorporated into national and worldwide socioeconomic policies are problems without short-term solutions. These problems are part of the world guidelines of research on the world sustainability.

It has become necessary to create new technical and political foundations to put pressure on the structures of current economic models that relate to increasing privatization of world biomes. This "surge in privatization" contradicts the conception of management of access to the planet's wealth in the long term. Religion, politics, and science, also worried about the future of its existence, have incorporated environmental issues in their agendas without breaking with the economic processes. A new international division of environmental enterprises has taken form in articulation with the interests of NGOs, important actors in the construction of worldwide citizenship, especially in Amazonia. The presence of these actors in Amazonia has been tense and contradictory. Therefore, the need has arisen to construct political instruments to keep the democratic system committed to new world political, social, ethnic, and ecological contracts reaffirming the geopolitical importance of the countries from the African, Asian, and Latin America continents. This book incorporates new elements into this framework, reaffirming the importance of Amazonia at this world juncture.

The sustainability paradigm proposes to construct a network of global integration through sustainable structures. This presents a new political perspective: the world's economic restructuring. The struggle against the social inequality will incorporate new socioeconomic tendencies into public policies and the market. It also enhances the possibility of the opposite: market and the national state creating new tools to increase social inequality. The degree of friction and asymmetry between these two processes is highly dependent on the evolution of the economic and political powers of the richest countries, the worldwide integration of basic public policies-health, education, and culture-and the politicization and organization of worldwide public opinion. The role of science and technology in the confrontation of "economic crisis vs. sustainability," embedded in the process of developing models of economic sustainability, is still under construction.

This complex socioeconomic universe has been designed and consolidated into the public policy of science and technology and integrated into Amazonia's cultural characteristics and production matrix. This region is being projected as the primary world center for sustainable development over the next decade.

However, its full opening to the predatory economic market and the structural changes to federal laws made by the current Brazilian government complicate the policies on environmental protection and climate change in this region. These actions from the Brazilian government also risk jeopardizing the possibility of Brazil becoming a developed country through the sustainable use of this region.

There is a fascist dimension of the current Brazilian government that potentiates the rapid ecological destruction of Amazonia

There are signs that the international community has already begun to discuss the first political steps to make Amazonia into common heritage of mankind. The implementation and unfolding of this issue are still unpredictable.

More radically, Professor Stephen Walt of Harvard University published on May 08, 2019 an article in *Foreign Policy* magazine asking: Who will Save the Amazon (and How)? Stephen proposes invading Brazil to stop the ecological destruction of the region. International pressure to protect Amazonia is ongoing. Global warming is the main trigger of this type of political action that is in the process of rapid dissemination around the world. The protection and development of Amazonia, without reifying it, is a challenge for many generations of young Brazilians. A tenacious fight against human stupidity and the financial market's greed.

This book will allow the reader a consistent understanding of the heuristic range of the concept of sustainability and its nexus with Amazonia and worldwide processes.

Through this book, the authors honor the Prussian scientist Alexander-von-Humboldt (1769–1859) for his brilliant works on the earth sciences. We also honor the photographer and activist Claudia Andujar (1931 – ....), Swiss-Brazilian, for her long struggle to protect the indigenous peoples of Amazonia.

We thank Professor Bruce Patrick Osborne for the competent grammatical review of this book. Special thanks to Michelle Smith, Divya Vasudevan, and to Editorial Team of Peter Lang Publishing.

Manaus, November 25, 2019
Marcílio de Freitas

References

Barifouse, R. (2019). 'Dia Mundial do Meio Ambiente: 68% das áreas de proteção e indígenas da Amazônia estão ameaçadas, diz estudo (World Environmental Day: 68% of Amazonia's protection and indigeous areas are threatened, says study)', *BBC News*, 5 June, <https:// www.bbc.com/ portuguese/ brasil-48504317>, accessed 22 July 2019.

McGrath, M. (2019). 'Deforestation: Tropical tree losses persist at high levels', *BBC News*, 25 April, <https://www.bbc.com/news/science-environment-48037913>, accessed 10 July 2019.

UN News. (2019). 'World is on notice' as major UN report shows one million species face extinction', *UN News*, 6 May, <https://news.un.org/en/story/2019/05/1037941>, accessed 15 August 2019.

Walt, S. M. (2019). 'Who will save the Amazon (and how)?', *FP-Foreign Policy—The Magazine*, 5 August, p. 1, <https:// foreignpolicy.com/2019/08/05/who-will-invade-brazil-to-save-the-amazon>, accessed 28 August 2019.

CHAPTER ONE

Amazonia: World Central Issue

1.1 Amazonia: Worldwide Material and Symbolic Representations

Amazonia is strategic to Brazil and mankind. It raises a number of important issues to the world, with emphasis on the following: its participation in the construction of a new aesthetic concept for mankind, its sustainable development as the world's largest open living library, as a strategic space belonging to Brazil and the world, as a means to renew the planet, as the planet's thermostat, and as the planet's climate stability mechanism. These meanings of Amazonia's material and symbolic representations need to be understood by mankind so that we value nature and life at all times.

The possibility of the planet's ecological instability reaffirms Amazonia's worldwide importance. At this time, climate change is its main "driving force." Amazonia's cultural and environmental complexity qualifies and problematizes its inclusion in the world's political and economic processes. Nature and culture, territories and peoples, economy and technological innovations, environmental services, and sustainable development are categories that drive the cycles of its physical and political existence. Its exploitation by predatory capitalist forces and its laboratory-like condition for great international scientific experiments have generated contradictions and controversies that create new world political tensions and diplomatic agendas. The emergence of sustainability has added new elements to this framework.

The invention of new industrial matrices, and the integration of communication networks with new forms of financial market organization and the technical occupations are indications of globalization processes. New world geopolitics and links between education, science, religion, and the financial market have radically changed our relationship with nature and society. The incorporation of sustainability within this framework and in people's fictional imaginary has enhanced the creation of new tendencies and forms of social organization.

These broad issues direct the globalization of sustainability towards the economic, religious, and scientific foundations of the civilizing process, generating a new political and ethical centrality. This centrality promotes and appoints local enterprises, based on an ethical system installed on the structural revaluation performed on the basis of notions of value and rights. Value and the right to life; the intrinsic value of nature and environmental services; culture, nature, and symbolic value; individual rights and environmental preservation; collective rights; and international relations, among others, constitute controversial themes that underpin projects, research programs, public policies, and international agreements. Giving new meaning to the foundations, explanatory meanings, and operative mechanisms for the notion of value also requires reviewing approaches and evaluations of human beliefs and desires. In this context, three classic categories, truth-value, utility-value, and beauty-value (Latour and Lépinay, 2008) have been replaced by life-value, mankind-value, culture-value, and universe-value, more complex categories and with a greater heuristic range (Freitas et al., 2017). This new conformation of mankind enhances the sustainability ethic and Amazonia.

The globalization of hypocrisy and the political barbarity have fortified the ethic's importance, considering sustainability as a universal paradigm. Its incorporation as a civilizing foundation demands changes in the relationships between man and nature, between nations, between consumers and the market, and between technical training institutions and national state organizations. Developed countries are redefining their productive and occupational matrices, assuming the paradigm of sustainability and its links with public policies and ecological issues. For this reason, national states are seeking to integrate science and technology with the market in an ethical perspective that values social promotion. This to require structural changes in teacher training in all fields of knowledge.

The sustainability ethic embraces different themes and programs designed for the production and construction of human life and of public policies, generating contradictions and ruptures between ecocentric and anthropocentric conceptions. This historical framework has weakened the public consensus based on sustainability's regulatory guides and standardizations. It is now necessary to institute principles and criteria that guide the organization of methodologies and priorities for individual and collective classifications (Routley, 2007), from a perspective that considers culture to be imbricated in nature, and that it reaffirms the basic relations of this sustainability's new ethic. An ethics guided by the brief period it has been associated with human needs and, simultaneously, by the extended period it has been articulated to the permanence of mankind and the planet (Meunier and Freitas, 2005). The fusion of mankind's history to the history of matter enhances this ethical dimension according to which man does not behave as nature's owner thus proposing to modify, internally, the predominant morals.

This new framework from the twenty-first century enhances social promotion and the protection of nature. National states and societies exert key roles in the proposed technical and judicial regulations, as well as in its political legitimatization. These changes are producing a new understanding of the concepts of economic development and citizenship. From the standpoint of social solidarity, its main causes are the structural changes in education, science, and technology. Ecology, climate change, and sustainable development gain prominence in this new world scenario.

Brazil as the primary environmental power in the twenty-first century and Amazonia as the primary world reference for sustainable development are central to this process, which raises new challenges in science education because of the planet's environmental degradation. Today, there is a worldwide focus on new technologies related to the following themes: clean development mechanisms; environmental policies associated with climate change; management processes dealing with environmental education and culture; scientific and technological mechanisms applied to the biogeochemical cycles and environmental services; institutional strategies and methodologies applied to the preservation of landscapes; processes of land use, climate, and integrated management of ecosystems; and methodologies to use and integrate meteorological networks and water resources. Sustainable development is a key element in this process, as it proposes new forms of organization and relations between societies, national states, and man and nature. Five new commitments from the twenty-first century are required to improve societies and mankind: a worldwide political contract; democracy as a universal political system; a new worldwide contract on nature; a new social contract; and a new contract on ethics.

A new contract on nature has been discussed by presidents of national states in various international forums. The planet's social and ecological stability is the central focus of this contract, which also has Amazonia as its primary reference (Freitas and Silva Freitas, 2013). This new contract on nature, with sustainability as its paradigm, has highlighted contradictions in the capitalist regimen.

The central question can be put as follows: How to accomplish economic development concurrently with social inclusion and the protection of nature? Economic systems are integrating research on novel industrial matrices and sustainability through networks and technological platforms. The models of development based on a polluting industrial matrix and on the use of fossil fuels contribute massively toward fast worldwide ecological degradation and exploitation. These models are in the process of extinction. They are being redimensioned and adapted to sustainability's new requirements. However, these changes are implemented at a slower rate than the impacts of climate change. Amazonia is a key element in this process. The construction of a market that values the preservation of natural environments also constitutes a new dimension of world economic processes.

But, Amazonia is also not exempt from the deleterious effects of climate change. The atmospheric modeling legitimized by the Intergovernmental Panel on Climate Change (IPCC, 2007) predicts that the planet's warming process also presents the feasible possibility of savanization of Amazonian biomes, directly impacting the eight countries that make up Pan-American Amazonia, and more than 250 endogenous cultures. This thermodynamic scenario impacts the region's biogeochemical cycles, endangering the survival of the largest tropical rainforest on the planet. Global climate change models need to be improved. Again, Amazonia is a key element in this process.

The world's environmental crisis, caused by predatory development, has resulted in a worldwide synergy for the preservation of natural resources, including soils, waters, and the earth's atmosphere. It has also contributed to the creation of new educational matrices committed to mankind's future and the struggle against social inequality. Even so, the financial market's pragmatic interests continue to prevail through intensive privatization and generating unlimited profit, and political alienation.

Uncontrolled population growth, environmental issues, religious fundamentalism, political extremism, massive immigrations, and growing social and economic inequality are problems with a direct impact on new development models and the quality of life.

World population is growing rapidly: one billion in 1830, two billion in 1930, three billion in 1960, four billion in 1975, five billion in 1990, six billion in 2000, seven billion in 2010, eight billion by 2025, and more than 9.5 billion by 2050. This shows mankind's demographic evolution in the last two centuries. Further, by 2050, 800 million people will be living under nutritionally deficient conditions

(Collomb, 2000), unless public policies change radically. The construction of a world food security policy centered on sustainability is a political challenge.

The paradigm of sustainable development gives historical prominence to Amazonia. If present trends continue, there will be insufficient natural resources for the world population's basic needs in 2070. A new contract on nature is therefore necessary. A set of commitments institutionalized by governments and incorporated into national and international public policies could ensure the technical instruments required to guarantee the planet's socioecological stability. The central axis of this contract (Freitas and Silva Freitas, 2014) is the preservation of the human species, an issue articulated by science education, ecology, and Amazonia via the processes of climate change. There is national and international consensus regarding Amazonia's importance to Brazil and to the world.

Amazonia is in South America. It is characterized by high temperatures, humidity, and heavy rainfall. It covers parts of Brazil, Peru, Ecuador, Bolivia, Colombia, Venezuela, Suriname, Guyana, and French Guiana and occupies approximately 6.5 million km^2 of which 4.5–5 million km^2 are forests. This region supports the contiguous area of the greatest social diversity and biodiversity in the world and is home to one-third of the rain forests of the world and one-fifth of the earth's surface fresh water. Amazonia also plays an important role in the mechanics, thermodynamics, and chemistry stabilities of the world's atmospheric processes. Brazilian Amazonia is formed by the states of Amazonas, Acre, Pará, Amapá, Roraima, Rondônia, Tocantins, Maranhão, and Mato Grosso. These nine Brazilian states are home to a population of approximately 30 million, which corresponds to 0.35% of the world's population. There are 163 different indigenous peoples, for a total population of 342,000 or 47% of the Brazilian indigenous population. About 22,000 isolated communities live within its forests.

Brazilian Amazonia covers a total of 4,987,247 km^2, 58% of the total area of Brazil and 40% of South America (5% of the earth's surface) (Silva, 2013). Approximately, 3.5–4 million km^2 is forest without significant anthropogenic disturbance. Brazilian Amazonia has approximately 75,000 km of navigable rivers, a fleet of 350,000 boats, and 11,280 km of borders with seven neighboring countries. In addition, there are 12 million hectares of wetlands and 150 million hectares of protected forests in federal and state conservation units (data from 2011). It plays an important role in the planet's climate and thermodynamic stabilities. Brazil is ranked first in the world in terms of diversity of plants, fish, fresh water, and mammals; second for amphibians; and third for reptiles. It possesses 55,000 different vegetable species (22% of all plant species) and 524 different species of mammals, 517 of amphibians, 1,622 of birds, 486 of reptiles, 3,000 of fish, 10–15 million insects, and millions of microorganisms. The majority of Brazil's patrimony is located in Amazonia, further emphasizing the importance of its insertion in world economic processes (Cruvinel, 2000). Scientific literature also confirms that scientists are aware of less than 10% of all existing biodiversity on earth. It has been stated that 40% of current medications in modern medicine were developed from natural sources; for example, 25% from plants, 12% from microorganisms, and 3% from animals. Furthermore, a third of the most prescribed medications worldwide comes from these sources. If anticancer drugs and antibiotics are considered separately, this percentile increases to approximately 70% (Calixto, 2000), which reaffirms the geopolitical and economic importance of Amazonia.

There have been a number of phytogenetic accomplishments in Ducke Forest Reserve, a preservation area (100 km^2) located close to Amazonas State capital Manaus. Researchers from Brazil's National Institute for Research in Amazonia (INPA), in Manaus, verified the existence of 5,000 individual trees and 1,200 tree species in this Reserve (Ribeiro et al., 1999). This is equal to the total number of species in Europe, reaffirming the great biological diversity of that area, where new species are still being discovered.

Through a refined prospecting, Hans ter Steegel et al. (2016) made a study using 530,025 collections between 1707 and 2015, identifying 11,676 tree species in Amazonia, distributed in 1,225 genera and 140 families. These studies are strategic to the development of a bioeconomy that is in urgent need of new taxonomic and floristic studies in a region with an estimated 16,000 tree species. Certainly, this great living library has answers to many of mankind's problems.

According to Kinver (2017), there are 60,065 species of trees in the world. Therefore, Amazonia is home to 19% of the world's tree species. It is home to the world's largest genetic bank on solid land or the largest botanical garden on the planet. Its connections with economic processes need to be better measured because of its importance for mankind's future.

In the words of Barata (2012): "the Amazonian forest has registered 2,000 medicinal species used by the local population as medicines, in addition to 1,250 aromatic species producing essential oils. However, only three aromatic species are part of Amazonia's export and trade agenda: cumaru beans, copaíba and rosewood essential oil." In short, this is a complex natural laboratory open to biotechnological innovations.

Research also indicates that Amazonian forests have 350 tons of biomass per hectare and produce annually 7.5 tons of vegetable litter (branches and leaves) per hectare, one of the largest world sources of renewable biomass on a solid surface. According to Antony (1997), in the forest on the Anavilhanas Archipelago, in Central Amazonia, which is subject to periodic flooding, a population of microbes with 116,409 individuals per m^2 was found in a superficial layer 10 cm deep.

Recent studies also reveal the existence of approximately 300 species of trees with a diameter of more than 10 cm per hectare in Brazilian Amazonian forests, exceeding the total number of species in Europe. The productive chains associated with bioindustry and with the chemistry of natural products need to be developed in the region.

About 20% of the Amazonian rainforest has been destroyed since 1970. The large economic projects installed in the region and biomass burning are the main actors in these predatory processes. Large gas and particle emissions from burning have a strong impact on its biomes and the composition of its atmosphere. Change in the cloud formation process, modification of cycles of various chemical elements such as ozone, important to the forest's stability, and the impact on the interaction between the electromagnetic radiation, light, and the forest affecting the entire photosynthetic chain are perverse effects of biomass burning in this region (Artaxo et al., 2005). Several ongoing research programs seek to understand and measure the range and the regional and planetary impacts of these processes, particularly on the greenhouse effect and climate change.

These features of Amazonia present challenges to forest engineering, to basic sciences, and particularly to biology, physics, chemistry, meteorology, and to engineering of new materials making possible new forms of management, and the production of new methodologies and sustainable products in the tropical rainforest basin.

The Brazilian government's political inability to propose forms of public-private partnership to exploit Amazonia's natural wealth, maintaining the forest preserved in a sustainable way, has sacrificed several generations of Brazilians. Financing Amazonia's economic and social development requires high investments, about US$1 trillion during ten consecutive years, in strategic Amazonian projects. The origin and application of these financial resources constitute a political action that needs to be executed, with public support, in the face of possible opposition from the hegemonic political interests of Brazil's south and southeast regions, mainly.

The Amazonian rainforest basin presents itself as a "water world." Its social and economic processes, its history and myths, geography, productive arrangements, and culture are driven by the cycles of nature permeated by the cycles of water and energy. Animal and plant life in Amazonia is inseparable from the cycles of nature. The meteorological sciences, agroecology, naval engineering, tropical medicine, anthropology, sociology of science, pharmacology, tropical technologies with emphasis on fish farming, information and communication technology, food technologies, ecological mining, design, and ecotourism, among others, are areas of science and technology essential to its sustainable development. Through research, innovation, and development programs; the integration and sustainable socioeconomic use of the Pan-Amazonian water basin is urgently needed. This is another Amazonian challenge.

The Amazonian region is crossed by the Amazonas River, which drains more than 7 million km^2 of land and has an average annual outflow of approximately 176,000 m^3/sec (176 million L/s). This makes it the world's largest river by volume of water, approximately four times the volume of the Congo River in Africa (second largest) and ten times the volume of the Mississippi River. In the dry season, the flows of Amazonas River into the sea at about 100,000 m^3/sec and at more than 300,000 m^3/sec in the flooding season (Sioli, 1991). The Amazonas River is also the longest and widest on the planet. It is about 6,992 km long and 8–10 km wide in the periods of low water and up to 50 km wide, in flat regions, in the flooding season. It can be up to 100 meters deep.

Based on the RADAM (Radar Project of Amazonia) inventory and other reference sources, Junk (1993) estimated that 20–25% of Amazonia's territory is periodically flooded. The high rainfall and the relief of the region favor this phenomenon, which covers about 1 million km^2 of its biomes. Junk's studies (1989) show that it is possible to measure the impacts of this phenomenon on the structures and processes that control the distribution of plants and animals, the primary and secondary production, and the nutrient cycles, among other important factors for the stability of these wetland forests.

The term "wetland forests" is used by Junk to refer to all types of forest subject to irregular, seasonal, or long-term flooding. Since Pre-Columbian times, this vast region has been inhabited by thousands of riverside residents who practice family farming and have livestock, causing only minor environmental impacts, and extract from nature only the products needed for their survival and trade with local organizations. Nowadays, this picture is in the process of major change due to the heavy pressure exerted by large logging companies, farmers, cattle ranchers, and large-scale fishing. These flooded forests play an important role in controlling primary production, carbon stock, and various biogeochemical cycles in the region, which is home to more than 1,000 species of trees (Junk and Piedade, 2010). Clearly, they, too, have connections with the control of climate change exerted by Amazonia. Its conservation and sustainable management and development is a challenge for Brazil.

Recent geological evidence also indicates the existence of an underground river about 6,900 km long, under the Amazonas River, at a depth of 4,000 meters. This underground river has a flow of 3 million liters (3,000 m^3/s). The two water courses flow in the same direction-from west to east-but possess different physical behaviors (Pimentel, 2013). Several international geological studies seek to understand the characteristics of this complex river basin.

The Amazonian basin has low demographic density and one of highest rainfall indices on the planet, with an average of 2,200 mm per year (1 mm of rainfall corresponds to 1 liter of water per square meter). This represents an annual total volume of water of 12×10^{12} m^3 (12 quadrillion liters), resulting in the world's largest rainforest (Salati et al., 1983). The region has more than 1,000 rivers forming the hydrological network necessary for its ecological and social integration.

Evapotranspiration is an important phenomenon for the thermodynamic stability of plants. Through this process, the leaves of each tree in Amazonia release about 300 to 1,000 liters of water per day into the atmosphere. This immense amount of steam rises to the atmosphere's upper layers forming the so-called flying rivers, which have a strong impact on regional and continental atmospheric processes. Researchers (Pinedo-Vasquez et al., 2013; Nobre, 2014) state that "In Amazonia, there are 5.5 million square kilometers occupied by native forests, with approximately 400 billion trees of the most varied sizes." "We did the count, which was also independently verified, and emerged the incredible number of 20 billion tons (or 20 trillion liters) of water that are produced every day by the trees in the Amazonian basin."

Therefore, there are the "flying rivers" in the Amazonian atmosphere, a complex hydrographical basin on its solid surface, and large hydrological reservoirs in its subsoil, not yet properly dimensioned. Understanding nature's engineering operation here presents a great challenge for future generations. The connections between these natural water courses and the energy transport processes in Amazonia involve several worldwide ongoing interdisciplinary research programs.

The energy moves the splendor of animal and plant life in the Amazonian basin. The intensity of solar energy in this region is 400 calories per square centimeter per second. One calorie is the amount of heat required to raise the temperature of 1 gram of water by 1 degree Celsius. It is estimated that 80–90% of this energy is used in the forest evapotranspiration process and 10–20% for air heating (sensible heat). There are days in which the temperature increases by up to 30 degrees for heights 10 km above the ground. In the forest on dry land, the steam originates from transpiration (70%) and from rain's evaporation intercepted by the forest canopy (30%). The Amazonia and Congo basins and the area around Borneo are typically tropical, important to earth's ecological stability and efficient in the absorption of solar energy and its redistribution via atmosphere (Crutzen and Andrae, 1990). Recent studies project that the humidity conversion process (via rain) in Amazonia's atmosphere liberates heat equivalent to approximately 400 million megawatts. This energy is essential to Amazonia's maintenance and to the thermodynamic stability of global atmospheric processes. These surveys are ongoing.

Amazonia's energy and waters are the physical "fuels" of its biological and cultural diversity. Interventions in its environments by Amazonian traditional populations have increased its biological diversity over time. After the arrival of man in this region thousands of years ago, three factors have emerged and contributed to the multiplication of its heterogeneity. First, the hunting or selective fishing of large animals, birds, and fish; second, fire, which has resulted in the opening of small clearings; and the introduction of exotic species such as dogs, oxen, cultivated plants, and the domestication of native species (Brown Júnior and Freitas, 2002). The habits, relationships, and movements of its fauna and flora in its biomes as well as its physical interactions with external environments also amplified its biological diversity. Finally, the continuous renewal of its cycles of nature and the circulation of its peoples in its natural gardens also contributed to the construction of its rainforests and rivers. Therefore, the sustainable development of Amazonia is also a commitment and a civic debt of the national state to its original populations that have bequeathed us this fantastic physical and cultural heritage.

The break with important international scientific and technological collaborations associated with several Brazilian research programs towards better understanding of Amazonian ecosystems functioning and use is another of the Brazilian government's mistakes. The destructive impact of this political action will affect several national institutions and generations of young Brazilians. Once again, political barbarism has triumphed over wisdom. Concentrated and long-term research, sophisticated studies, important social and economic projects applied to region, specialized technical training programs, and collaborations of a universal nature, among others, are in the process of extinction. A particular example is the destruction of the Amazon Fund Program, as presented in Chapter 7. Brazil still does not have enough science and technology for the full development of Amazonia in sustainable patterns. The integration and sustainable development of Pan-American Amazonia require alliances with traditional populations and international collaboration in knowledge from various fields.

The scientific and technological education of the 163 indigenous peoples from Brazilian Amazonia is another challenge. The organization of their academic curriculum requires the following innovations: reaffirmation of the education of indigenous teachers committed to science education integrated to the complexity of the region; an interdisciplinary pedagogical project emphasizing the indigenous teachers' participation in teaching the subjects of this program; incorporation of themes touching on philosophy, the natural sciences, and the social sciences; inclusion of approaches and technical innovations that enable an interface of Western science and traditional knowledge; formation of networks promoting their scientific inclusion; formation of vocational technology platforms; and the access of the students to contemporary educational issues, strengthening their cultural formation and confrontation with predatory capitalism.

Teacher training programs for the indigenous peoples with a view to making Amazonia sustainable should be guided by the following thematic axes: mathematization and geometrization of nature's processes; history and philosophy of science and indigenous education; research methodology in science education for indigenous teachers; psychopedagogical models applied to science education for indigenous teachers; instrumentalization for the training of indigenous teachers in science education; training of indigenous teachers in the models of sustainable development; indigenous education, communication, and contemporaneity; and engineering for the rainforest, health technicians, among other less important topics (Freitas and Pires, 2012). The Indigenous University of Amazonia is still a distant reality. This institution would generate many innovations in public policies and sustainable development of the region and the world. Its implementation is also a challenge to the region's leaders.

How to develop Amazonia? This is a key issue both regionally and nationally. The sustainable development of Amazonia requires a set of operational initiatives among which we highlight: the building of a national political consensus; heavy investments in research, innovation, and development; increased support of the regional research and technology institutions; guide the basic science and tropical technological innovations to regional public policies; induce new tropical technologies; development of new differentiated industrial and fiscal policies; appropriate infrastructure and logistics; new regulatory framework for sustainable ecological and economic zoning; participant planning including all social and economic actors; and consolidation of national and international scientific partnerships. Sustainable development inserts new elements into Amazonia's future.

In contrast, the federal government's incentive of the possession and illegal use of indigenous lands by prospectors and loggers, and its threat to revise consolidated indigenous rights are another perverse legacy from the new president of Brazil. The growing expansion of agribusiness in Amazonia and the recent authorization by the Brazilian government of 51 different types of highly toxic pesticides also constitute threats to its indigenous peoples. Agribusiness continues to advance through Amazonia, creating poverty for its populations, deforesting and contaminating its biomes.

Without exception, all of the major projects that have been proposed for Amazonia are guided by predatory capitalist development. Proposals for a new territorial division and environmental cartography, economic de-structuring, exacerbated privatization, frequent illegal occupation of extensive territories, and unrestricted access to the large economic groups that use or collect nature's resources on a large scale, among others, has generated a lot of tension and will contribute to the growth of poverty among their populations and the irreversible destruction of its ecosystems. Amazonia's social poverty and environmental destruction are historical and political monuments to this type of capitalism, as shown in Chapter 6.

The use of organized and accumulated knowledge about Amazonia and the humid tropics enables its sustainable development without cutting down a single tree of its forests or polluting its rivers. This should be a development policy guideline for Amazonia. Partnership with their traditional peoples will accelerate the use of this strategic guideline for this region.

Sustainability makes an original contribution to this situation. Today, Brazilian society is acting as a political brake against the disastrous actions of Brazil's new president to Amazonia. The international trade and economic pressure against the Brazilian government is also a key element in Amazonia's protection.

Major world newspapers have published their critical opinions about Amazonia's destruction. *The Economist* says that "The Amazon is perilously close to the tipping-point. Brazil has the power to save earth's greatest rainforest-or destroy it." This same magazine calls all presidents to take up a political position against Bolsonaro's environmental program; he is considered the world's worst enemy of the environment (*The Economist*, 2019). *The New York Times* also recorded the fast spread of deforestation in Amazonia since Brazil's new president, Jair Bolsonaro, took over and his government reduced efforts to fight illegal logging, ranching, and mining (Casado and Londoño, 2019). In his article, Stéphen Rostain, Director of Research at CNRS, published in *Le Monde* magazine, criticizes the world's passivity in the face of the burning in Amazonia. Rostain asks: "Is Amazonia burning? Yes, but apparently not enough to move powerful policymakers and funders, the archaeologist warns" (Le Monde, 2019a). On September 2019, "over 200 investors representing some U$16.2 trillion under management called on companies to do their part in halting the destruction of the world's largest tropical rainforest" (Slattery, 2019). This issue needs to be immediately incorporated into the industrialized countries' diplomatic agendas.

Protests against the destruction of Amazonia are rapidly multiplying worldwide. Some of the untimely actions of the Brazilian government harmful to Amazonia have been annulled by collegiate decisions of Brazilian Superior Courts. However, this has not prevented the destruction of many programs of protection and social promotion in Amazonia. In this sense, Amazonia's political future is also very uncertain. This news is spreading rapidly worldwide.

It is important to remember that from 2003 to 2016, the Brazilian government of the Workers' Party implemented important social programs in Amazonia. In this period, Programs such as "Light and Water for All," "Family Allowance,"

"Zero Hunger," "My House – My Life," and "More Doctors," among others, have all enabled the social inclusion of more than 4 million people in the region (data 2016). Geographical characteristics, large distances, low population density, climate regimes, and heavy rainfall in the region have been obstacles to making these Programs more efficient. These social initiatives have been important to boost sustainable development in Amazonia's isolated regions. Brazil's new president has closed down most of the Programs. This new state of affairs raises the following question: How to build a social and economic sustainability network for the social promotion of the population and Amazonia's environmental protection in a context of extreme political adversity?

Supporting the inclusion of Amazonia's protection in international political forums, in new protocols, in contracts, and in global public policies will be an effective counterpoint to its ongoing destruction. These actions are necessary and urgent.

Contradictorily, Amazonia's political and economic importance grows as its role in the ecological stability of the planet and mankind is reaffirmed and policies for its economic development are weakened. This situation favors new forms of colonialism in the region, recreated by scientific, political, and business leaders. New tensions are emerging between the Region, the Nation, and the World.

Sustainability and culture, sustainability and the sacred, sustainability and protected spaces, sustainability and nature and the city, sustainability and the economy, and sustainability and public policies are dimensions essential to development models in Amazonia.

Amazonia also plays a special role in the essential processes of ensuring the chemical stability of the earth's atmosphere. Experts speculate it contributes, on a regional and international scale, to the levels of carbon dioxide (the main greenhouse gas), nitric oxide, and nitrogen dioxide, key agents responsible for the degree of atmosphere oxidation, and nitrous oxide gas, approximately 200 times more harmful than carbon dioxide (Keller et al., 1983). The degree of importance of the first two nitrogenized gases to the chemical stability of the atmosphere and of the other two to climate stability is a complex problem and still subject to scientific research. The Intergovernmental Panel on Climate Change stated that in 2014, approximately 15.98 billion tons of carbon dioxide were emitted into the earth's atmosphere. Amazonian ecosystems behave like a gigantic vacuum cleaner, absorbing, for photosynthetic effect, 250–500 million tons of this gas per year (Gash et al., 1996). Higuchi (2007) estimates, based on an average of 160 tons of carbon per hectare, that Amazonia's ecosystems store 90 billion tons of carbon, 13% of the total carbon stored in earth's atmosphere. Uncontrolled deforestation, therefore, has an immediate impact on the growth of the greenhouse effect.

Amazonia is not the lung of the world. The oxygen it produces is counterbalanced by what it consumes. There is a small excess released that has no major impact on the total volume of oxygen present in the earth's atmosphere.

The ATTO Program, Amazonia Tall Tower Observatory, was inaugurated in the State of Amazonas in August 2015. This program is the result of a scientific collaboration between the Max Planck Institute, Germany, the National Institute for Research in Amazonia, and the State University of Amazonas, among others. It possesses a Tower of 325 meters high, fully instrumentalized to measure the interaction between atmospheric processes and the Amazonian rainforest. These measures will enable an estimate of the degree of the Amazonia's participation in the planet's climate, chemical, and thermodynamic stabilities. In the long term, it is intended to measure the Amazonia's participation in climate change, creating elements to plan the use and occupation of its ecosystems. The Tower is the planet's largest free laboratory for atmospheric studies and will have a period of continuous use of 20 to 30 years. The development of this Research Program has been compromised due to the lack of interest of the federal government in continuing it. The impacts of its interruption have not yet been fully measured, but several scientific projects associated to this Program have already been paralyzed.

The environmental monitoring of Amazonia by Brazil's National Institute of Space Research (INPE) confirms that the development policy of Brazil's new president for this region has caused the fast growth of its deforestation. This deforestation increased 88% from 2018 to 2019 (INPE, 2019). This is a tragedy for Brazil and mankind. It is a political problem in need of an urgent solution. Brazilian society and the international community need to intensify political pressure on the Brazilian government to guarantee protection of Amazonia.

In August 2019, Brazil's president fired the Director of INPE, due to the disclosure of the increasing rates of deforestation in Amazonia (Quierati, 2019). INPE has been monitoring deforestation in Amazonia since the 1980s. Yet, Jair Bolsonaro claims the Institute has inflated these deforestation rates, and that INPE has an ulterior motive in disclosing them. He also decided that from that time on this technical information would be of strategic interest to the Brazilian State. Nontransparent control of this information by the State, the technical disqualification of INPE, and the disrespect to its Director, Ricardo Galvão, an eminent Brazilian scientist, intensify the uncertainties about the protection of Amazonia. These actions are another political aberration of this government.

In December 2019, Professor Ricardo Galvão was chosen by Nature Research Journal as one of the ten people who mattered in Science in 2019, for his action in defense of Amazonia.

The Amazonian region's economic potential grows together with its importance in balancing the world's ecological stability. Its economic production rose to US$70 billion in 2012, the result of more than 600 industrial enterprises, predominantly in electro-electronics, computer science and technology, motorcycles and bicycles, pharmaceutical products and cosmetics, and mining exporting to more than 80 countries. The economic values of the environmental services rendered to the world by the Amazonian ecosystems are estimated at about US$3 trillion per year.

New relations between science, religion, regional development, Amazonia, and sustainability present complex problems to mankind. Currently, Brazilian Amazonia is undergoing an economic and political cycle tending toward irreversible ecological destruction (Freitas and Silva Freitas, 2018a,b). Its physical and cultural heritage needs to be protected. Based on this assumption, it is necessary to disseminate its importance and to develop it in sustainable form.

The Brazilian government's incompetence to implement a modern industrial policy contextualized to Amazonia's cultural and ecological potential is an obstacle to its economic development. This question can be formulated as follows: How to implement a sustainable industrial policy in Amazonia exploring its natural resources?

Amazonia's future has sustainable development seated in its multinaturalism and the technology as reference. Education, science, social and economic equity, religion, and world processes are key elements in this challenge. However, since January 2019 Amazonia has been devastated by more than 75 thousand fires, which represents 84% more than last year (Le Monde, 2019b). Its destruction must be stopped. As Stephen Walt (2019) puts it: "Who will Save the Amazon (and How)?"

The political positions of the presidents of Brazil, the United States of America, and China on the environmental protection of our planet are worrying. Their environmental policies put many uncertainties for mankind. The Catholic Church has always been very active in Amazonia since the arrival of the first Europeans in this region. Anticipating the national state's disastrous political actions in the region, Pope Francis led the Organization of the Special Assembly of the Synod of Bishops to reflect on Amazonia's main problems. In October 2019, in Rome, this religious institution defined its new guidelines and operational actions in this region of such importance to the world. From an ecumenical perspective, the globalization of Amazonia is another great challenge for institutions committed to building world peace and ensuring the planet's environmental protection.

The UN and UNESCO, responsible for world security and defense of the physical and cultural heritage of the world's populations, should also be able to foresee and prepare for the disastrous and frightening political scheme unfolding in Amazonia. These institutions should urgently convene specific Assemblies to discuss protective mechanisms for the region. Religion, institutions for world security, and culture need to be aligned in the defense of Amazonia. Mankind's future is at stake.

The relationship between the globalization of environmental issues, Amazonia, and worldwide uncertainties will be analyzed in the next chapter.

References

Antony, L. M. K. (1997). 'Abundância e Distribuição Vertical da Fauna do Solo de Ecossistemas Amazônicos Naturais e Modificados (Abundance and vertical distribution of soil fauna of modified and natural Amazonian ecosystems)'. In N. Higuchi, J. B. Ferraz, L. M. Antony, F. Luizão, and Biot, Y. (eds.), *Projeto Bionte: Biomassa e Nutrientes Florestais—Relatório Final (Bionte project: biomass and forest nutrients—final report)*, pp. 225–31. Manaus, Amazonas: Editora do Inpa-MCT.

Artaxo, P., Gatt, L. V., Leal, A. M. C., Longo, K. M., Freitas, S. R., Lara, L. L., Pauliquevisi, T. M., Procópio, A. S., and Rizzo, L. V. (2005). 'Química atmosférica na Amazônia: A floresta e as emissões de queimadas controlando a composição da atmosfera amazônica (Atmospheric Chemistry in Amazonia: The forest and the biomass burning emissions controlling the composition of the Amazonian atmosphere)', *Acta Amazonica*, 35 (2), 185–96.

Barata, L. E. S. (2012). 'A economia verde – Amazônia (The green economy)', *Ciência e Cultura, São Paulo, On-line version*, 64 (3).

BBC News. (2017). 'O que são os 'rios voadores' que distribuem a água da Amazônia? (What are the "flying rivers" that distribute Amazonia's water?)', By *BBC News*, 1 September, https://g1.globo.com/natureza/noticia/o-que-sao-os-rios-voadores-que-distribuem-a-agua-da-amazonia.ghtml. Accessed 22 January 2019.

Brown Júnior, K., and Freitas, A. V. L. (2002). 'Diversidade biológica no Alto Juruá: avaliação, causas e manutenção (Biological diversity in the High Juruá: evaluation, causes and maintenance)'. In M. C. Cunha, and Almeida M. B. (eds.), *Enciclopédia da Floresta (Encyclopedia of forest)*, pp. 39–40. São Paulo, SP: Editora Companhia das Letras.

Calixto, J. B. (2000). 'Biodiversidade como fonte de medicamentos (Biodiversity as source of drugs)', *Ciência Hoje*, 28 (167), 36–43.

Casado, L., and Londoño, E. (2019). 'Under Brazil's far-right leader, Amazon protections slashed and forests fall', *The New York Times*, 28 July, p. 3.

Collomb, P. (2000). 'Quelle sécurité alimentaire pour les pays en développement em 2050? In J. Bindé (ed.), *Les clés du XXI siècle*, pp. 129–130. Paris: UNESCO-Seuil.

Crutzen, P. J., and Andrae, M. O. (1990). 'Biomass burning in the tropics: impact on atmospheric chemistry and biogeochemical cycles', *Science*, 250, 1669–78.

Cruvinel, T. (2000). 'Brasil, um país megaecológico (Brazil, a Megaecological Country)', *Jornal A Crítica*, 5 May, p. A5, Manaus.

Freitas, M., and Pires, J. (2012). 'Formation scientifique des peuples amérindiens d'Amazonie et durabilité: proposition et engagements', *La Recherche en Éducation*, 7, 66–86.

Freitas, M., and Silva Freitas, M. C. (2013). *Sustainability; Man-Amazonia-World.* Baltimore: America Star Books.

Freitas, M., and Silva Freitas, M. C. (2014). 'Fragments of utopias from twenty-first century and Amazonia: projections and controversies', *Innovation: The European Journal of Social Science Research*, 27 (3), 275–94.

Freitas, M., and Silva Freitas, M. C. (2018a). 'Science, religion and Amazonia: education for sustainability', *International Journal of Environmental & Science Education-IJESE*, 13 (3), 275–93.

Freitas, M., and Silva Freitas, M. C. (2018b). 'Regional development for sustainability in Amazonia: controversies and challenges', *Geography, Environment, Sustainability-GES*, 11 (4), 112–31.

Freitas, M., Silva Freitas, M. C., Ioris, A., and Castro Júnior, W. E. (eds.) (2017). *Amazônia (Amazonia).* Lisboa: Chiado Editora.

Gash, J., Nobre, C., Roberts, J., and Victoria, R. (eds.) (1996). *Amazonian deforestation and climate*, pp. 549–76. Chichester: John Wiley & Sons.

Higuchi, N. (2007). *'Carbono da vegetação do Estado do Amazonas (Document on vegetation carbon from the Amazonas State)'*, pp. 1–16. Manaus: Printed notes.

INPE. (2019). 'Alertas do DETER sobre desmatamento e degradação ambiental na Amazônia em junho somam 2.072,03 km^2 (DETER's alert about deforestation and environmental degradation in Amazonia in June cover 2,072.03 km^2)', *INPE News*, 10 July, <http://www.obt.inpe.br/OBT/noticias/alertas-do-deter-na-amazonia-em-junho-somam-2-072-03-km2>, accessed 12 July 2019.

IPCC. (February 2007). 'Climate change 2007: the physical science basis; summary for policymakers'. Working Group I, Paris, France.

Junk, W. J. (1989). 'Flood tolerance and tree distribution in central Amazonian floodplains'. In L. B. Holm-Nielsen, I. C. Nielsen, and Balslev H. (eds.), *Tropical forests: botanical dynamics, speciation and diversity*, pp. 47–64. New York: Academic Press.

Junk, W. J. (1993). 'Wetlands of tropical South America'. In D. Whigham, S. Hejny, and Dykyjova D. (eds.), *Wetlands of the world*, pp. 679–739. Dordrecht: Junk Publications.

Junk, W. J., and Piedade, M. T. F. (2010). 'An introduction to South American wetland forests: distribution, definitions and general characterization'. In W. J. Junk, M. T. F. Piedade, F. Wittmann, J. Schöngart, and Parolin P. (eds.), *Amazonian floodplain forests – ecophysiology, biodiversity and sustainable management*, pp. 3–25. Dordrecht/Heidelberg/London/New York: Springer.

Keller, M., Goreau, T. J., Kaplan, W. A., and McElroy, M. B. (1983). 'Production of nitrous oxide and consumption of methane by forest soils", *Geophysical Research Letters*, 10 (12), 1156–59.

Kinver, M. (2017). 'World is home to 60,000 tree species', *BBC News*, 5 April, <https://www.bbc.com/news/science-environment-39492977>, accessed 10 June 2019.

Latour, B., and Lépinay, V. A. (2008). *L'économie, science des intérêts passionnés*, pp. 36–8. Paris: Éditions La Découverte.

Le Monde. (2019a). 'Pourquoi la forêt amazonienne brûle-t-elle?', *Le Monde*, 22 May, p. 1, Paris.

Le Monde. (2019b). 'Incendies: l'Amazonie paie la politique du président Brésilien', *Le Monde*, 23 August, p. 2, Paris.

Meunier, O., and Freitas, M. (2005). 'Culturas, técnicas, educação e ambiente: uma abordagem histórica do desenvolvimento sustentável (Cultures, techniques, education and environment: a historical approach to sustainable development)'. In M. Freitas (ed.), *Amazônia; a natureza dos problemas e os problemas da natureza (Amazonia: the nature of the problems and the problems of nature)*, pp. 106–141. Manaus: Editora da Universidade Federal do Amazonas.

Nobre, A. D. (2014). *'O Futuro Climático da Amazônia—Relatório de Avaliação Científica (Amazonia's climate future – Scientific assessment report)'*, p. 17. São José dos Campos: Edição ARA, CCST-INPE/INPA/WWF.

Pimentel, E. T. (2013). 'Estrutura Termal da Crosta na Região Amazônica com Enfoque no Movimento de Águas Subterrâneas nas Áreas de Bacias Sedimentares (Thermal structure of the crust in the Amazonian region with focus on the movement of underground waters in the areas of sedimentary basins)'. Ph.D. thesis, Federal National Observatory, Rio de Janeiro.

Pinedo-Vasquez, M., Sanjayan, M., Patterson, T., and Sukhdev, P. (26 March 2013). 'Amazon's 'invisible flying rivers''. In *BBC Future – BBC.com. In this film these researchers reveal how this water is carried across Latin America.*

Quierati, L. (2019) 'Demissão de Diretor do INPE é oficializada no Diário Oficial da União (Resignation of INPE Director is made official in the Official Gazette of Union)', *Uol News*, 7 August, <https://noticias.uol.com.br/politica/ultimas-noticias/ 2019/08/07/demissão-de-diretor-do-inpe-e-publicada-no-diario-oficial-da-uniao.htm? cmpid=copiaecola>, accessed 5 August 2019.

Ribeiro, J. E. L. S., Hopkins, M. J. G., Vicentini, A., Sothers, C. A., Costa, M. A. S., Souza, J. M., Martins, M. A. D., Lohmann, L. H. P., Assunção, L. G., Pereira, P. A. G. L., Silva, E. C., Mesquita, C. F., Procópio, M. R., and Lilian, C. (eds.) (1999). *Flora da Reserva Ducke. Guia de identificação das plantas vasculares de uma floresta de terra-firme na Amazônia Central (Flora from Ducke reserve, identification guide of vascular plants of the a mainland forest in Central Amazonia)*. Manaus: Midas Printing LTd., Inpa – DFID.

Routley, R. S. (2007). 'A-t-on besoin d'une nouvelle éthique, dúne éthique environnementale?'. In H.-S. Afeissa (ed.), *Éthique de l'environnement—Nature, valeur, respect*, pp. 31–49. Paris: Librairie philoophique J. Vrin.

Salati, E., Junk, W., Shubart, H., and Engrácia, A. (1983). *Amazônia: desenvolvimento, integração e ecologia (Amazonia: development, integration and ecology)*, p. 126. São Paulo: Editora Brasiliense.

Silva, M. C. (2013). *Metamorfoses da Amazônia (Metamorphoses of Amazonia).* Manaus: Editora Valer.

Sioli, H. (1991). *Amazônia: fundamentos da ecologia da maior região de florestas tropicais (Amazonia: foundations of the ecology of largest region of tropical forests)*, p. 68. Petrópolis, RJ: Editora Vozes.

Slattery, G. (2019) 'As Amazon burns, 230 big investors call on firms to protect world's rainforests', *Reuters News*, 18 September, https://www.reuters.com/article/us-amazon-investors/as-amazon-burns-230-big-investors-call-on-firms-to-protect-worlds-rainforests-idUSKBN1W32T3, accessed 12 October 2019.

Steegel, H., Rens, W. V., Dairon, C.-L., Daniel, S., Alexandre, A., Sylvia, M. O., Nigel, P., Peter, M. J., Rafael, P. S., and Vitor, H. F. G. (May-Aug 2016). 'A descoberta da flora arbórea da Amazônia com uma lista atualizada de todas as taxas arbóreos conhecidas (The discovery of the Amazonian tree flora with an updated checklist of all known tree taxa)', *Boletim do Museu Paraense Emílio Goeldi Série ciências naturais*, 11 (2), 231–61.

The Economist. (2019). 'Deathwatch for the Amazon – the threat of runaway deforestation', *British Magazine*, 1 August, p. 1.

Walt, S. M. (2019). 'Who will save the Amazon (and how)?', *FP-Foreign Policy—The magazine*, 5 August, p. 1, <https:// foreignpolicy.com/2019/08/05/who-will-invade-brazil-to-save-the-amazon>, accessed 28 August 2019

CHAPTER TWO

The Globalization of Environmental Issues and Amazonia

2.1 Introduction

Deadlocks that guide the current political processes, and have the ecological sustainability paradigm as foundation, are analyzed in this chapter. Controversies and uncertainties of the twenty-first century, surrounding scientific processes and educational policies on a worldwide scale, are also dealt with. Special attention is given to ecological issue that permeates economic and scientific mega-processes. Finally, the insertion of Amazonia in this world process is analyzed.

2.2 The Globalization of Environmental Issues

Industrialization burns coal, oil, and gas in order to generate energy to fulfil social needs. Western countries, China, and Japan lead this enterprise responsible for rapid environmental degradation: a serious threat to the future of mankind.

The economic and political establishment, under pressure from the agroindustrial sector, has failed to present proposals that inhibit the deterioration of environmental issues. This sector is the most dependent on fossil fuels- mineral coal, petroleum, and natural gas. It is also the principal sector responsible for soil, water, and air pollution, with impacts on all public policies. This same group, in conjunction with scientific and political groups positioned in strategic regions, develops a marketing methodology to add economic value to animal and vegetable products. Pharmacology, biotechnology, and new materials industries seem to want "to gobble up the forests." Never have so many patents been registered in such a short time frame.

The development of large economic conglomerates is still partly supported by poor countries. Pillaging in the colonial period; intense exploitation of natural resources at degrading costs- fossil, minerals, forest, marine, and the genetic heritage which continues to threaten several populations in these countries; unequal world trade conditions that favor the exportation of goods without reference to the social impacts generated in the process of extraction or production; intellectual expropriation and use of traditional native knowledge by agroindustry and biotechnology, without payment of the intellectual property rights; loss of large stretches of land and water sources to intensive agriculture for the purpose of production for export, posing a serious threat to food security and to the cultural values of the native populations; the effect of worsening air pollution contributing to the greenhouse effect and to the destruction of ozone layer; and finally, the construction of nuclear weapons and chemical substances that put all mankind at risk are perverse examples of these economic conglomerates (Lefetey, 2002).

The inventory of the world's biodiversity has already identified about 1.75 million species on the earth, of which 4,500 are animals; 10,000 are birds; 1,500 are amphibians and reptiles; 22,000 are fish; 270,000 are plants; 70,000 are fungi; 5,000 are viruses; 4,000 are bacteria; 400,000 are invertebrates, not including insects; and 960,000 are insects, of which about 600,000 are beetles (Dallmeier, 2000). Specialists speculate that these projections represent less than 10% of the actual number of species on the planet, most of which are in the oceans and tropical regions, more than 50% of them residing in Pan-American Amazonia, Central Africa, Southeast Asia, and in areas of Australia. After four years work in the forests of the Brazilian Amazonia, scientists from the Emilio Goeldi Museum, based in Belém and within the administrative scope of the Ministry of Science and Technology, announced on February, 19, 2014, the discovery of Amazonia's 169 new species of fauna and flora: 14 plants and 155 animals. Their work is part of an online platform dedicated to the Inventory of Planetary Biodiversity that counts with 20 participating world institutions (Goeldi Museum, 2014).

Amazonia has an important position in the research and consolidation of the natural sciences and in the global spread of Western culture. Two important moments highlight the importance of Amazonia in the history of science. First, Charles-Marie De La Condamine's participation in the French scientific expedition in Pan-American Amazonia from 1735 to 1744. During his descent along the Amazonas River, La Condamine made astronomical observations associated to his geographical positions; he studied the habits and customs of the populations of the region; the identification, distribution, and uses of new animal and plant species, biogeography, and the use of new medicinal plants in pharmacology and industry; the cycles of nature- the water cycle, the solar cycle, and the life cycle of plants and animals; fluid mechanics- in particular the behavior of rivers, their floods and ebbs; the tropical biomes' optical properties- brightness, reflection, refraction, and light absorption; the movements of the stars in the tropics; and the use of the rubber tree, curare, and quinine (La Condamine, 1992). These chemical products had great impact on English pharmacology.

The main purpose of these studies was to prove the validity of the Physical Models of Mechanics- Motion Theory presented by Isaac Newton (1642–1727) in his book *The Mathematical Principles of Natural Philosophy* published in England, in 1687 (Newton, 2010). This book is an emblematic reference for the history of Western science. Newton introduced new approaches and physical and mathematical language to explain the movement of objects. He also made a synthesis of the knowledge accumulated by the physical sciences, created new physical-mathematics solutions for the movement of objects in resistive and non-resistive media, and universal articulations between earthly and celestial physical phenomena.

A similar level of importance must be given to the development of biology with the scientific contributions made by Alfred Russell Wallace (1823–1913) during his expedition in Amazonia from 1848 to 1852. His main book *A Narrative of Travels on the Amazon and Rio Negro, With an Account of the Native Tribes, and Observations on the Climate, Geology, and Natural History of the Amazon Valley*, published in London in 1853, presents important structural contributions for the natural sciences (Wallace, 1853). His studies in Amazonia resulted in the publication of this book and 11 scientific articles. In 1853, during his expedition to Borneo, Wallace wrote and published the article "On the Law Which Has Regulated the Introduction of New Species." (Wallace, 1855). These works were important references for the construction and legitimation of the theory of evolution of the species by natural selection. In 1859, Charles Darwin (1809–1882) published this theory in its definitive format (Darwin, 2009).

The discoveries of Wallace and other researchers in this expedition to Amazonia had a great impact on the development of biology. Just to illustrate: Henry Walter Bates, another scientist on this expedition, during his stay in Amazonia from 1848 to 1859, collected 14,712 different mammalian specimens, birds, reptiles, and insects, 8,000 of them new to science (Bates, 2009). English museums and pharmacology were revitalized with these discoveries from Amazonia. European scientific expeditions in Amazonia have played an important role in the formation of social thinking about the region. They also contributed to the construction by the rest of the world of a new interpretation and representation.

The theory of the evolution of the species is a structural reference in the history of science. It has established a fruitful dialogue between the natural sciences and the human sciences, unfolding in the incorporation of the molecular explanation of the various branches of biology: cellular biology, virology, immunology, physiology, neurology, endocrinology, and also ecology. New links have been created between science, nature, religion, and the market.

Amazonia has been presented as a "living library," strategic to bioindustry and to transnational groups. Curare, rubber, and quinine are examples of Amazonian products, of immense use commercially, pirated and appropriated by the Europeans during the colonization process in the eighteenth century. The same type of piracy continues to happen on a large scale in this region. This reaffirms the importance of the institutionalization of an integrated development policy for its complex structures through concessions and privatizations with different regulations and participations.

At an ever increasing pace, the disorganized occupation and inadequate use of the tropical regions have caused a forest loss at a rate of approximately 0.8% to 2% per year. As part of this projection, specialists estimate a loss of up to 16 million genetic populations per year, in other words, loss of one population every two seconds. As far as species are concerned, a loss of 27,000 species per year is estimated, or one species every 20 minutes (Purvis and Hector, 2000). It is estimated, on a global scale, that 654 plant species and 484 animal species, of which 58 species are mammals and 115 species are birds, have disappeared since initial contact in the seventeenth century.

The rate of deterioration of the world's biodiversity is fast increasing. Inadequate use of soil and water; commercial overexploitation of some species; introduction of predatory species into certain ecosystems; growing soil, water, and air pollution; intensification of agribusiness; reordering of territories; and global climate change are its main causes.

Other factors, also, have contributed to this process, among which are accelerated demographic growth; economic development policies unadapted and unintegrated to the actual regional environment; the nonregulation of access to right of use of natural resources; and the insufficiency of scientific knowledge about regional and world ecological dynamics (Lévêque, 1997). Large extensions of cutting and burning in tropical regions play an important role in the deterioration of this troubling situation. Amazonia has strong insertion in this world environmental panel.

Defense of Nature has given rise to speeches, which come from different ideological spectra, from the most radical to amply permeable. In the forefront, there are NGOs that defend strict protection of nature, the creation of more natural reserves, more rigor and a faster response in the application of the law and the environmental codes, better use of renewable resources, and monitoring of animals and plants species. In opposition to these NGOs there are several transnational economic groups, allied to the liberal governments and to regional economic actors, who argue in favor of more flexible and wider use of natural resources. The new Brazilian environmental policy is inserted within this liberal orientation which insists on imprisoning the world and Amazonia in the shadows of political ignorance.

Still disorganized, uninformed, and excluded from decision-making processes of these two tendencies, the populations of the tropical region are unable to protect and enjoy nature's assets developed by them and by the generations that preceded them.

The process of capital accumulation ranges since the global circulation of virtual merchandise to the natural capital, including soils, waterways, the atmosphere, and all their components, whether biological or no. The planetary dimensions of ecological issues and the environmental services rendered by the planet's several ecosystems, in particular by Pan-American Amazonia, Southeast Asia, and Central Africa, constitute an array of economic interests that articulate regions, countries, and continents. Currently, the environmental issue is part of the national projects of all countries.

Costanza et al. (1997) says that the "(…) Economic projections for environmental services, on a planetary scale, already surpass the total sum of the Gross Domestic Products from eight richer countries". Contradictorily, the populations from these three tropical regions are submitted to an increasing process of pauperization. Today, these areas, which jointly have more than 50% of the world's biodiversity, have become great laboratories for ecological research activities. The vastness of the possibilities that these regions should "imprison" as yet undeciphered scientific aspects continues to be one of its nexus with economic groups, which have the hegemony of high-tech technologies.

Current tendencies have shown intense and increasing involvement of NGOs in environmental issues in these three tropical regions. Education and environmental monitoring, development of new technologies and technical and social assistance to rural populations, conservation and management of natural resources, environmental inspection and mobilization of public opinion, and ecotourism and ecological research are themes that make up the agenda of many world institutions. Modernity has discovered that the "tropical world" constitutes an important source in the global dynamics of genetic storage and in climate change. Its cultural and environmental destruction has a strong impact on productive arrangements and world societies.

Most research policies on biodiversity in these regions are organized along four major axes: relations between nature and culture, processes of mass and energy transport in natural ecosystems, preservation and restoration of biodiversity, and economic and social values of biodiversity.

In general, these programs propose integrated management for ecosystems; conservation of genetic resources and biogeochemical cycles; conservation and recovery of fauna resources; recovery of degraded areas through agroforestry systems and other alternatives; the management and use of secondary vegetation; the management of wood and non-wood products; the integrity of hydrographic basins, maintenance of aquatic ecosystems and their resources; water quality; limnology- biology of aquatic organisms and fish ecology; ethnobiology and social representations; and the fisheries and fish culture and landscape ecology (Freitas and Silva Freitas, 2013a).

Gradually, science and the world environmental movement have executed a process of denaturalization of the last ecological frontiers by assigning them value. The complete disjunction between scientific disclosure, application of techniques, the occupational matrix, and social demands on these economic frontiers give rise to several contradictions imbricated in its economic processes. The low technical level of its productive arrangements conspires against its sustainable development.

The evolution of science and technology has enabled the emergence of an increasingly technological industrial matrix. The incorporation of new products and management processes into public policies improved the quality of life of people and made possible new forms of organization of cities and societies. Relations between countries, people, and the environment have changed. On the other hand, they have caused environmental degradation and social inequality that has put mankind's and the planet's future at risk.

A prevailing tendency in environmental ethics studies highlights that the central problem is not limited to the invention of a new morality, but proposes the construction of strategies that enable the application of the dominant moral schemes to the new world ecological order: environmental pollution, increasing process of pauperization, accelerated destruction of the world's biodiversity, the greenhouse effect, and the deployment of nuclear power plants. This same trend considers that the Western philosophical, scientific, and religious tradition has a theoretical arsenal, powerful enough to deal with the problems posed by the ecological crisis.

A second tendency, however, has considered that, far from providing the elements of a solution, this very tradition is part of the problem (Larrère and Larrère, 1997). A point of view that reinforces this latest trend is one that signals that Western tradition shows that there are no perfect solutions that are impervious to criticism, to the great social, political, and scientific questions put before mankind. Its representatives deny the perfect organization of a society that is harmonious, free from domination and without conflicts, and affirm that political decisions should be considered as solutions to increasingly complex social problems and subject to criticism and review (Albert, 1987).

In this century, the fictional and mechanical language of techniques is deeply tensioned and intertwined with the divine and sensitive representations of mankind. It is also definitively incorporated into a new world political framework permeated by exacerbated social inequality and the possibility of the ecological destabilization of the planet.

Without nature there is no technique, and without technique there is no bioindustry.

In this sense, the set of environmental issues presented constitutes epistemological elements for several research projects, and it is the axis of different multilateral actions in the contemporary world. The heuristic range of these research programs surpasses barriers of sciences, involving political constructs and wider universal commitments on the use of nature. Among the various possibilities, these commitments could result in an optimistic alternative for building a more equitable and solidary world.

Emphasis to the impacts of the greenhouse effect and to climate change. The consequences of this effect caused by the intensification of the concentration of carbon dioxide in the earth's atmosphere, produced by the burning of fossil fuels, management and deforestation have already been incorporated into government diplomatic agendas. The correlation of this effect with climate stability has social and environmental impacts in all time and space dimensions.

Burning the forest releases many contaminants into the atmosphere, hazardous to the population's health. Children are most affected by this type of pollution (Carrillo et al., 2019). Together, the four countries with the highest carbon dioxide emissions in 2014, at a total of about 62% of global carbon dioxide emission, were China (30%), the United States of America (15%), the European Union (10%), and India (7%) (Olivier et al., 2015). From 1990 to 2014, total aggregate greenhouse gas emissions of carbon dioxide with emissions/removals from land use, land-use change, and deforestation decreased by 15.8%, from 18.98 to 15.98 billion tons carbon dioxide (UN, 2016). It is important to emphasize that there are still great uncertainties in the methodologies of greenhouse gas emissions estimates as well as in carbon dioxide emissions from China. Amazonia is a key region in this worldwide process.[1] Current lack of control over the growth of deforestation and burning in this region is a worldwide concern.

Another global controversy refers to the growing appropriation of organizational processes and biological differentiation, on a molecular level, by the biotechnology industry. This issue projects a future with new relations between nature and development. Small but significant differences in the molecular structures of biological organisms have contributed toward the intensification of research projects on genetic evolution. Full control over animal and plant cloning processes, the development of transgenic products, and the ample domain of the processes of construction of life are targets, which guide the corporate actions of several transnational conglomerates in this sector. Possibility of transforming life into commodities has already been incorporated by markets and important financial centers. Man's desire to control his own evolution through the ability not only to conserve species in its integrity, but also to improve it, transforming it to suit the project of its creator is ongoing.

Studies on the borderline between chemistry and biology, and between biophysics and biochemistry mediated by ethics are highly complex philosophical and historical issues. The Principle of Responsibility, as formulated by Jonas, calls for the need to make the effects of man's actions compatible with the existence of a genuinely human life on the earth (Jonas, 1990).

In a certain sense, genetic engineering has a similar role for the financial market in the twenty-first century to that of the atomization of nature at the beginning of the twentieth century. Advances in the microscopic readings and measurement techniques, major availability of natural resources, and the increasing appropriation and expropriation of indigenous peoples' knowledge by the biotechnology industry will advance and accelerate the consolidation of this market that already turns over some trillions of dollars. And worse still, as in the past, scientific and technological processes described as revolutionary have increased social inequalities among people, creating new forms of colonization.

By year 2000, the United States of America and Canada had dominated research and innovation in biotechnological processes: 76% and 7% of the biotech industries were North American and Canadian, respectively (Bilan Du Monde, 2002). According to the 2015 OECD Report, the United Sates ranks first in the number of biotech firms, PCT patent applications, and biomedical treatment

1. The current concentration of 370 parts of carbon dioxide per a million parts of air (370 ppmv) shall increase to about 550 ppmv in 2100, for a scenario of lesser emission or above 830 ppmv for a scenario of higher emission. These projections reaffirm the greenhouse effect's gravity in the world climatic stability (Olivier Jos et al., 2015).

approvals. It has 11,367 biotech firms followed by 2,831 in Spain and 1,950 in France. They are followed by Korea, Germany, the United Kingdom, with Japan, Mexico, New Zealand, and Belgium rounding out the top 10 (Phillips, 2019).

Sustainability and bioindustry have technical and symbolic relations. This productive arrangement is fundamental for sustainable development. It also contributes to universal imagery committed to the preservation of nature. The tendency to the planet's continuous ecological destruction has induced and reaffirmed the importance of this type of development. Amazonia is an immense laboratory for bioindustry, though its insertion in this market is still in the process of technical and legal regulation.

Systematic and irreversible destruction of the world's primary ecosystems[2] by the developing countries as well as the intensification of the use of highly polluting technological matrices constitute a counterpoint to sustainable development. This type of development gained political legitimacy in the UN Report "Our Common Future," approved by the Assembly of the United Nations, in August 1987 (Brundtland et al., 1987). But, the search for worldwide political consensus is ongoing. European countries lead this process that raises new uncertainties and contradictions for the technical world.

2.3 Uncertainties and Controversies of the Twenty-First Century

The severe downward economic trend in developing countries projected for this century tend to reinforce the rise of populist, authoritarian governments. Increasing difficulty of access to public policies by the majority of their populations favors the eruption of political scenarios with unforeseeable consequences for environmental policies favorable to maintenance of the planet's ecological stability.

World Bank projections estimate that an investment of 5% of overall expenses by the governments of developing countries would reduce the number of undernourished children from current 166 million to 94 million in 2030 globally. However, if the current political situation persists, in excess of 500 million people may not have safe and continuous access to basic food security, and 130 million children in preschool age will live in subnutritional conditions in many urban outlying districts in nondeveloped countries in this same year. This multilateral institution also projects the need for the governments of developing countries to obey the

2. In this context, the term "primary" designates the natural configurations, specific of soils, waters, and atmosphere comprising an ecosystem.

law, adopt transparency in their political actions, eliminate corruption, and respect and protect human rights as preconditions to achieving sustainable development. It further emphasizes that the failure of public policies in developing countries is directly attributable to their governments (World Bank, 2011; Lin, 2013).

However, this same institution omits and fails account for the fact that the poverty and appalling social indicators of these countries were also caused and institutionalized, in the past and the present, through economic and political programs, which articulate the interests of the governing elites of rich countries and this multilateral institution associated to big business. It also glosses over the political and economic blackmail to which this institution (the World Bank) and the IMF subject peoples of outlying regions. Based on more complex and realistic methodologies, critical experts also question these statistics and project more dramatic scenarios for mankind's future. And they insist on the need for structural changes in world political and economic systems, which permeate current civilizing processes.

Political manipulation, media control, institutionalized corruption, spurious agreements, and absence of organized civil society and democracy in the vast majority of these developing countries have contributed to crystallizing this situation.

A more consistent and inclusive civilizing substratum is needed to mold and modulate the foundations of global economic processes committed to cultural and racial diversity and to the fight against social inequality.

In this context, constructing and planning new educational policies, at all levels, acquires special meaning. Tendencies of educational systems that adopt the "Principle of Competitiveness" and "Financial Imperative" as basic assumptions need to be modified, adopting a tendency permeated by "equity," instead.[3]

The reference is to the principles presented by Martin Carnoy (1999) who has classified educational reforms in the context of the globalization of culture, in the following manner: the reforms based on competitiveness whose paradigm is the evolution of the demand for qualification *vis-à-vis* the needs of the national and international labor market, organization for the production of academic results, and professional competence. The central axes of this kind of reform are decentralization, and the national management of educational norms and educational means. The reforms underwritten by financial imperatives are based on public deficit reduction and the transfer of state-controlled national resources to the private sector, including investments in education; it is imbued with a strong private-oriented

3. Equity is a notion that comprises persons, communities, peoples, and nations. A notion that is intertwined to social diversities and to right for self-determination. It potentializes the political and economic instruments necessary for each development model.

component and supports a reduction in public spending on education. And finally, the reforms based on fairness. Its are anchored to the assumption of equity and a quality education of. This tendency pays special attention to women, to minorities in addition to strengthening special education programs. This reform should also incorporate environmental education, with ecology as its central theme: ecology as a process of production and construction of life in its physical, artistic, and religious dimensions. Therefore, this educational reform will create links between critical education and the sustainability of people and the planet.

Free education and quality teaching at all levels of public education should be assumed as a collective conquest and a universal right. For this reason, it cannot be subjected to market regulation. Budgets for the public financing of education need to be increased immediately, in a differentiated way and establishing new international forums. This would force and fix targets so that countries can strengthen their educational systems in a form compatible with its needs and educational deficits. Environmental education must permeate all ongoing educational tendencies. The populism of Liberal governments and the increasing privatization of education make it difficult to implement public and critical educational policies. People without technical training have difficulty entering the labor market (Coile and Duggan, 2019).

Agreements, exchanges, and diplomatic agendas between developed and developing countries need to be recreated and revitalized. There is an urgent need to establish new contours and delimitations of global public debt by making available greater financial resources in order to fight global poverty and illiteracy. Prominence to the critical situation associated to the growing exploitation of child labor in poor countries and to the premise that increasing the number of children in school does not imply growth, in same proportion, in the number of educated, qualified children, capable of earning a reasonable wage in the exercise of their profession. In general, mobilization on behalf of universal education is totally detached from the labor issue (Schlemmer et al., 2002).

There is a consensual thesis that attributes co-responsibility to the agents of the globalization processes for the increasing deterioration of the world's ecological and sociocultural indicators. The creation of specific global audits on the financial capital that supports these processes becomes crucial. It proposes to establish taxes and specific fees[4] to facilitate the continuing flow of investments in support

4. In 1972, James Tobin, Nobel Prize in economics, proposed the collection of a tax on all the speculative transactions in the world finance markets. Tobin Tax, as it became known, in the amount of one-tenth of 1 %, would enable to raise funds equivalent to an annual sum of about US$ 166 billion, at that time. Enough value to eradicate the world's extreme poverty from that period until the twenty-first century. The indifference from hegemonic governments in creating the political conditions, necessary to operate an alternative proposal on the growing world social inequality, constitutes a worldwide political discomfort. Samuel Benchimol (1992) proposed the

of projects to be applied in regions with more precarious social indicators. The sophistication of current accounting methods makes it possible to implement global taxes coupled to the world networks and in real time. Although it is fragmented, this proposal would enable, at its limit, to construct reverse tendencies to the situation of world poverty, contributing to a more fraternal and just dialogue among the different peoples of the world.

World scenarios projected for the future are definitively contaminated by the "bioterrorist" threat and by political and religious intolerance. This framework amplifies the difficulties of diplomatic forums to mediate a stable and systemic solution for the end of wars, environmental degradation, and social inequalities. A new conception of world political organizations is needed. The crisis is systemic and global although local issues could provoke unexpected political scenarios.

September 11, 2001 exacerbated the crisis and has become a reference in the history of mankind projecting international terrorism as the stigma of this new Age.[5] Mankind's old ghosts have reappeared: the full control over individual, collective, and world security; the unpredictability of terrorist attacks; and the confrontation "East-West."

Terrorist movements have changed their strategies and methods. The rigid and hierarchized structures of these groups during the 1970s and 1980s have been replaced by horizontal network-shaped articulations, with mass extermination as its central objective. The developed countries are very vulnerable to terrorist attacks due to their organizational structures. Their large urban concentrations, mass transport systems and administration of essential services such as electricity and water production, and the communications networks have contributed to the stimulation of a confrontational discourse. This discourse has reinforced the conservative and reactionary thesis for the development of new technologies for military application to guarantee national and world safety. The proliferation and increasing possibility of using biological weapons in terrorist attacks scares civil populations and contributes to obstructing the construction of world peace (Delpech, 2002).

The USA air attack on Iraq on January 3, 2020 with the death of Iranian General Qassim Soleimani and several Iraqi military has intensified the confrontation between West and East. The diplomatic and military developments of this confrontation will have still unpredictable outcomes.

creation of a similar tax for the protection of biomes and the most endangered animal and plant species from the world.

5. It dates the terrorist attack that killed 3,025 people (projection not yet officially confirmed) in World Trade Center, New York, inaugurating a new Age of confrontation "East-West."

The expansion of the war industry induces new international conflicts. The tendency to growth in global military expenditures constitutes an obstacle for the consolidation of world peace. The fast growth of the USA's military budget[6] has reverberated in world geopolitics. Western European countries, following a period of deceleration in their military expenditures since 1998, have also increased this type of expenditure, prioritizing the reactivation of several military-oriented research projects. The contours of the war industry are very mobile; however, political scenarios have favored its expansion. The emergence of global public opinion more concerned with the world's peace and future has constituted one possibility to brake this belligerent tendency. Delpech (2002) states that China has also announced a 17% increase in its military expenditure.

The increase in the capacity of weapons of mass destruction is tragic. In addition to the great loss of human life, wars will contaminate the interconnected hydrological networks, and exterminate large numbers of animals and areas with vegetable cover. This non-sustainable scenario is a counterpoint to global environmental protection policies. According to Sigal Samuel (2019), "Scientists want to make it a war crime to damage the environment in a conflict." With this proposal, sustainability also puts new anti-war elements into political parliaments and diplomatic relations.

A report issued by the World Trade Organization (WTO) that evaluates trade of goods in 2015 and part of 2016 forecasts major uncertainties in this sector for the next few years. High subsidies for production, restrictive government policies, intensification of tensions among key sectors of the world's economy, increasing poverty, and the decrease in the use of computers and electrical and electronic products are some of the factors that have contributed toward worsening this situation. This fact causes economic and political insecurity and increases the possibility of ensuing serious impacts in the world's occupational matrix. The WTO also foresees an increase in global environmental destruction.

The World Summit on Sustainable Development alternately called "Rio+10" or Johannesburg 2002, held August 26 to September 4, 2002, is an important reference for the political tendencies of an important group of industrialized countries. These countries, led by the United States, Japan, Canada, Australia, and New Zealand, formed a movement whose objective was to transfer the responsibility

6. The military budget of United States of America in 2000 amounted to US$374 billion, about 1.83 higher than the sum total of military expenditures by other world's seven richest countries (Japan, Germany, France, United Kingdom, Italy, Canada, and Russia) in this same year. High investments by China and Russia in the development of new weapons have contributed to a new arms escalation. World peace is still under great threat.

for the international negotiations on sustainable development from the United Nations to the WTO. This tendency has become hegemonic.

The strategy has been to minimize the State's responsibilities for ecological protection, dividing them among private institutions. At the World Summit, the results agreed on by the 200 national representatives have proved to be dismal in view of the seriousness of the current ecological and social issues. This international forum has established the preservation of natural resources for future generations, prioritizing sustainable development.

In this forum, the following themes were discussed: eradication of poverty; the need to change current non-sustainable consumption and production patterns; protection and appropriate management of the sources of natural resources that constitute the basis for economic and social development; sustainable development for a globalized world; health and sustainable development; sustainable development for several places and regions (islands, Africa, Latin America and the Caribbean, Asia, and regions of Europe and the Pacific); forms and strategies for the establishment of sustainable development; and systems of institutional references for sustainable development.

In this same forum, many actions were presented as contributions to the stimulation and responsible institutional articulation toward sustainable development at local and national level (UN, 2002). Measures to protect earth and mankind's future have also been established, the most important of which are: to reduce by half the proportion of people without access to appropriate sanitation; to improve access to energy; to protect the seas and fish stocks; to minimize the harmful impact on humans and the environment; to increase aid to poor countries; and to eliminate aspects of globalization prejudicial to poor countries. The final text has also motivated trading without emphasizing that the WTO's rules exclude the global environmental agreements. This fact was considered a victory by environmentalist groups, which had feared that agreements such as the Environmental Protocols could be affected.

This same text presents the requirement that developed countries condition their aid to poor countries to a decrease in corruption and an increase in democracy; to developing a strategy to preserve natural resources for future generations; to reaffirming the Principle of Precaution as a basis for action in environmental protection whenever there is potential future damage to the world's ecosystem; and finally, a principle of common and differentiated responsibility, which reaffirms that all countries have to endeavor to save the planet, but that the developed countries need to provide the greater part of the funding (UN, 2002).

This meeting confirmed a tendency toward the political fragilization of such forums, reaffirming the deadlock in the construction of a collective and consensual solution to halt environmental degradation and poverty in the developing countries. Creation of codes of conduct, systems of technical and financial cooperation, and definition of common policies do not justify the absence of practical deliberations by this forum. There a need for preventive measures against accelerated global environmental and social degradation conflicts with the political interests and strategies of the majority of the governments of the developed countries. Clearly, the political and economic forces from hegemonic countries already have full control over these forums.

The resolution adopted by the General Assembly in September 25, 2015, "Transforming our world: the 2030 Agenda for Sustainable Development" is also an important reference for mankind (UN, 2015). This agenda extends the Millennium Development Goals and presents 17 Sustainable Development Goals, 169 targets, and sets deadlines and means of implementation for them. In general, these Goals of universal scope and significance have 2030 as their point of reference.

These Sustainable Development Goals are: End poverty in all its forms everywhere; End hunger, achieve food security and improved nutrition, and promote sustainable agriculture; Ensure healthy lives and promote well-being for all at all ages; Ensure inclusive and equitable quality education and promote lifelong learning opportunities for all; Achieve gender equality and empower all women and girls; Ensure availability and sustainable management of water and sanitation for all; Ensure access to affordable, reliable, sustainable, and clean energy for all; Promote inclusive, and sustainable economic growth, full and productive employment, and decent work for all; Build resilient infrastructure, promote inclusive and sustainable industrialization, and foster innovation; Reduce inequality within and among countries; Make cities and human settlements inclusive, safe, resilient, and sustainable; Ensure sustainable consumption and production patterns; Take urgent action to combat climate change and its impacts; Conserve and sustainably use the oceans, seas, and marine resources for sustainable development; Protect, restore, and promote sustainable use of terrestrial ecosystems, sustainably manage forests, combat desertification, and halt and reverse land degradation and halt biodiversity loss; Promote peaceful and inclusive societies for sustainable development, provide access to justice for all, and build effective, accountable, and inclusive institutions at all levels; and Strengthen the means of implementation and revitalize the global partnership for sustainable development.

Full UN control by a small group of countries makes it difficult to comply with this important agenda. Wars, social and economic inequalities, privatization of the world, environmental issues and climate change, political and religious intolerance, terrorism, lack of complete public policies accessible to all, and the absence of a political project for mankind are problems that reverberate in our daily lives and in society. New commitments and global contracts need to be discussed and institutionalized. This issue will be analyzed in the next chapter.

References

Albert, H. (1987). *La sociologie critique en question*, pp. 53–54. Paris: Press Universitaires de France.

Bates, W. H. (2009). *The naturalist on the River Amazons*. Oxford: John Beaufoy Pub.

Benchimol, S. (1992). *Amazônia; A guerra na floresta (Amazonia; the war in the forest)*. Manaus: Edua.

Bilan du Monde (2002). Paris: Le Monde.

Brundtland, G. H. (1987). *'Our common future – report of the world commission on environment and development'. Document presented in forty-second session – UN General Assembly*, New York, on 4 August 1987.

Carnoy, M. (1999). *Mondialisation et réforme de l'éducation: ce que les planificateurs doivent savoir*, pp. 41–51. Paris: UNESCO.

Carrillo, B., Branco, D. K., Trujillo, J. C., and Lima, J. E. (2019). 'The externalities of a deforestation control policy in infant health: evidence from Brazil', *Economic Development and Cultural Change*, 67 (2), 369–400.

Coile, C. C., and Duggan, M. G. (2019). 'When labor's lost: health, family life, incarceration, and education in a time of declining economic opportunity for low-skilled men', *Journal of Economic Perspectives*, 33 (2), 191–210.

Costanza, R., D'Arge, R., De Groot, R., Farber, S., Grasso, M., Hannon, B., Limburg, K., Naeem, S., O'Neill, R., Paruelo, J., Raskin, R. G., Sutton, P., and Van Den Bet, M. (1997). 'The value of the world´s ecosystem services and natural capital', *Nature*, 387, 253–60.

Dallmeier, F. (2000). 'Biodiversity: Earth's most important and most threatened asset', In Xunta de Galicia (ed.), *Reunión Internacional de Expertos en Educación Ambiental – nuevas propustas para la acción*, pp. 454–5. Santiago de Compostela: Editions UNESCO.

Darwin, C. (2009). *The origin of species*. Cambridge: Cambridge University Press.

Delpech, T. (2002). 'Politique du caos – L'autre face de la mondialisation', *La Republique des Idees*, 48, pp. 26–30. Paris: Seuil.

Freitas, M., and Silva Freitas, M. C. (2013a). *Sustainability; Man-Amazonia-World*. Baltimore: America Star Books.

Goeldi Museum. (2014). 'New species are discovered in Amazonia', *Museu Goeldi*, 28 February, < https:// www.museu-goeldi.br>, accessed 4 October 2018.

Jonas, H. (1990). *Le Principe Responsabilité*, pp. 31–56. Paris: Éditions Champs Flammarion.

La Condamine, C.-M. (1992). *Viagem pelo Amazonas, 1735–1745 (Journey through the Amazonas, 1735–1745)*. Rio de Janeiro: Editora Nova Fronteira.

Larrère, C., and Larrère, R. (1997). Du bom usage da la nature, pp. 292–293. Paris: Alto Aubier.

Lefetey, P. B. (2002). 'Dette financière contre dette écologique'. In D. Anne-Marie (ed.), *Les nouveaux utopistes du développement durable*, pp. 196–7. France: Éditions Autrement-Collection Mutations, France.

Lévêque, C. (1997). *La Biodiversité*, pp. 77–87. Paris: Presses Universitaires de France.

Lin, J. (2013). 'Industrial policy revisited: A new structural economics perspective', *Revue d'économie du développement*, 21, 55–78.

Newton, I. (2010). *The mathematical principles of natural philosophy*. Texas: Snowball Publishing.

Olivier Jos, G. J., Janssens-Maenhout, G., Muntean, M., and Peters, J. A. H. (2015). *'Trends in global CO_2 emissions: 2015 report; background studies'*, p. 16. The Netherlands: Netherlands Environmental Assessment Agency Institute for Environment and Sustainability of the European Commission's Joint Research Centre.

Phillips, T. (2019) 'The top countries for biotech firms and research', The balance, 28 January, <https:// www.thebalance.com/ranking-the-top-biotech-countries>, accessed 22 March 2019.

Purvis, A., and Hector, A. (May 2000). 'Getting the measure of biodiversity', *Nature*, 405(6783), 212–19.

Samuel, S. (2019) 'Scientists want to make it a war crime to damage the environment in a conflict', *Vox*, 26 July, https://www.vox.com/future-perfect/2019/7/26/8931699/environment-war-crime-scientists-nature-geneva-convention, accessed 12 August 2019.

Schlemmer, B., Martin, J. Y., Ahmed, A. S., and Waast, R. (2002). 'Le défi de La mondialisation: marchandisation ou partage des savoirs?', In J.-Y. Martin (ed.), *Developpement durable? Doctrines, pratiques, évaluations*, pp. 265–85. Paris: IRD Éditions.

UN. (2002). 'Plan of implantation of sustainable development, elaborated in the Meeting of Sustainable Development accomplished in Joanesburgo (online)', *UN* 2002, <http:// www.un.org>, accessed 20 June 2016.

UN. (2015). 'UN-General Assembly, Seventieth session, Transforming our world: the 2030 Agenda for Sustainable Development', *UN*, 25 September, United Nations, New York.

UN. (2016). 'UN-Framework Convention on Climate Change, National greenhouse gas inventory data to the period 1990–2014'. In *Framework convention on climate change*, p. 7. November 2016, United Nations, Marrakech.

Wallace, A. R. (1853). *A narrative of travels on the Rivers Amazon and Rio Negro, with an account of the native tribes, and observations on the climate, geology, and natural history of the Amazon valley*. London/New York/Melbourne: Ward, Lock, and CO.

Wallace, A. R. (September 1855). 'On the law which has regulated the introduction of new species', *Annals and Magazine of Natural History*, 16 (2nd Series).

World Bank. (2011). *World development report 2011: conflict, security, and development*. Washington, DC: World Bank.

CHAPTER THREE

The Utopias of New World Contracts and Amazonia

3.1 Introduction

The global mega-scenarios projected for this century are controversial and complex. The last century was stigmatized as marking the end of utopias. Specialists predict that ethical, economic, and political patterns that will emerge in the twenty-first century will echo and overlay the philosophical principles and the socio-economic scenarios prevalent last century. They claim that the structural demands of capitalism have generated cyclical crises and that they produce a continuous process of self-recycling, thus boosting social inequalities.

Studies also project that the future wars of mankind will not be limited to economic and ideological reasons, but be in defense of "civilizing frontiers" (Mattelart, 1996), and for the planet's environmental protection. These specialists insist that the "bellicose wave" led by the USA and the European countries against a set of countries called by the USA the "Axis of Evil"- currently Iraq, Afghanistan, Iran, Syria, and North Korea- will have an unpredictable denouement. The insertion of Palestine and Israel in this framework raises new uncertainties. The transformation of China into a great economic power has strengthened Eastern civilization and its insertion in political forums. Together, China and Russia have developed dialogues and diplomatic actions to build an equilibrium of political forces in this region.

The confrontation "East×West," the growth of social and economic inequalities, the multiplication of authoritarian political regimes, and the planet's ecological destruction are the main civilizing references of this century.

Theoretical surveys show that the foundations of the twenty-first century revolve around the following axes: construction of a worldwide political contract, centered on the confrontation "East×West" and the elimination of racism; crystallization of participatory democracy within a universal political system; implementation of a new natural contract, grounded in the possibility of the destruction of the planet and the extinction of the human species; consolidation of a social contract, reaffirming the universal rights of man; and affirmation of an ethical contract among nations, in response to the dilemmas and impasses delineated above, revealed and exacerbated by the globalization process.

These are the utopias of the twenty-first century. Utopia is a universal category. In certain cultural contexts, it can be seen as a reactionary counterpoint to the historical praxis, which proposes to change the "world," on a critical basis way and with the prospect of freeing man from his alienation and from capitalist exploitation. It also can be seen as referring to the solution of social problems, or as a virtual structure that considers idealistic intellectual abstractions. This analysis uses utopia as a methodological strategy to construct a consistent analytical text. Several utopias associated to modernity's relevant issues are analyzed.

Sustainability still constitutes the utopia of utopias. It is very dependent on ecology as a process of production, construction, and reproduction of life, in all its material and symbolic dimensions. History shows that capitalism is incapable heuristically of exploiting the planet economically while preserving its great biomes and populations. This reaffirms the importance of a new economic and political reordering of the capitalist system, and the need to reformulate the concepts of economic development and citizenship. Sustainability is a key element in this process. However, to fully attain sustainability new global contracts must be developed: contracts that will be widely effective for a long period of time, and which strengthen the distribution of wealth, the construction of peace, and better social and economic integration among countries. This study presents a brief conceptual description of the new utopias of the twenty-first century.

3.2 Political Utopia: New Worldwide Political Contract

The confrontation "East×West" involves different worldviews. Diplomatic dialogue between East and West has always been very fragile. Western culture's deadlocks have been enlarged by the growth of social and economic inequality between rich and poor. This issue conspires against the construction of peace and global solidarity.

The East's difficulty, in the majority of its countries, which is to construct a cumulative process and prioritize science and technology as instruments intrinsic to modernity, has hindered its dialogue with the West. Violence as a historical necessity of the central capitalist countries, added to China's political nonintegration into the international community, magnifies these uncertainties. The economic development achieved by China, Japan, and South Korea in the last two decades has changed this global picture (Lin, 2013). North Korea's success in the construction of ballistic missiles and India's space project adds new elements to the confrontation "East×West."

The political utopia which underpins a new worldwide political contract proposes to decentralize the economic and political powers of the developed countries. This issue presupposes the development of a new political centrality and a new time metric for economic and social processes. This political centrality is to be radiated from "regions" and a temporal metric rooted both in the brief time period of human needs and in the longer time period of planetary preservation. This situation requires giving new meaning to the concepts of citizenship and economic development (Freitas and Silva Freitas, 2013a). It also requires greater international cooperation for better regulation of the global financial market. The Financial Stability Board (FSB) is a key institution in this process.

Eric Helleiner (2011), in his contribution to the understanding of this global economic and financial scenario, explains the importance of the FSB in this process and highlights the G20's role in forming a new consensus on international financial market regulation. Helleiner highlights macro-prudential objectives to avoid the wider systemic financial risks, FSB efforts for more effective regulation of financial institutions, new laws encompassing the scope of the emerging financial and economic complexities, and greater regulation and stricter supervision of the parallel banking system, among other secondary ones. Experts do not perceive the financial system as a source of the economic and political processes that build world poverty and environmental degradation. They insist on a conception of exclusionary sustainability, as if there were one sustainability for the poor and another for the rich.

The growing financial flow from West to East puts new elements in this world economic process. Although preservation of the environment and poverty are not yet concerns central to the global institutions that control the production, circulation, and reproduction of capital; organized societies are demanding the inclusion of sustainability in public policies.

This means that the structural conditions are being built for a new framework that incorporates sustainability as the foundation of the civilizing processes. The multiplication and politicization of social and economic networks and the new forms of social groups organization contribute to the implementation of a political utopia, a new worldwide political contract.

This contract proposes to establish democracy as the universal political system, in which the economic processes and the legal framework are permeated by the new legal regulations and political determinations of the capitalist system. It also proposes the development of a new global economic and political order focused on elimination of the cult of racism and anthropocentrism. At this new juncture, it will be possible for each culture to have their own values and perspectives: love; religious belief; art; oral communication; technical knowledge; and social, economic, and political organizations, in different intensities and gradations. The democratic regime is still the best alternative, always with the prospect of satisfying basic needs and social promotion in a sustainable form.

3.3 Democracy and the Universal Political System

The irradiation of democracy is due to the expansion of Western culture and the assumption, in thesis, of the worldwide economic and legal framing of the capitalist system's political determinations. The guarantee of freedom and human dignity is still the democratic system's main reference; it will expand, rupturing the last worldwide authoritarian resistance. However, the real benefits of participatory democracy in developing countries will still be heavily dependent on political decisions in central countries.

Increasing economic and technical-scientific unevenness between rich and poor countries contributes to the prevalence of an asymmetrical globalization that includes the African continent and the majority of Asian and Latin American countries. This globalization has created an accelerated process of precariousness and pauperization in the populations of these countries. Currently, social inequality is induced by waves of consumerism and privatization, and is spread and crystalized in irreversible forms in large cities, regions, countries, and continents, putting the democratic system's credibility at risk. It challenges the modern state's capacity to guarantee the rights to all citizens.

Public policies involving health, education, housing, food, and labor have begun to collapse in developing countries. Access to justice and state protection of citizens from the predatory market are still at least partially unattainable in these countries.

The vulnerability of these countries' economic models to financial speculation, in accordance with the interests of the national and international economic elite, propagate instabilities in development projects, holding them hostage to large transnational corporations. Regional inequalities and intense internal competition for the state's financial resources amplify the political instability. The absence of democratic and popular national projects and the civil disorganization have allowed the colonizing actions of the oppressor to be used against the public institutions of these countries. Transformation of the world into a great financial market contributes to their political disarticulation.

Sustainability emerges as a possibility to protect even the most fragile populations and natural environments, in the present and the future. It follows that a new dimension of democracy will require a strong commitment to worldwide social and natural contracts; and it will strengthen the geopolitical importance of African, Asian, and Latin American countries.

In this century, democratic processes will develop the first definitive contours of worldwide citizenship while civilizing utopias. The search for this unity, in the face of the diversity and the plurality of world cultures, constitutes the main challenge of this political regime.

Specialists suggest the development of a new world economic order as the main presupposition leading to the crystallization of democracy in developing countries (Besley and Persson, 2019). The main focus would be the fight against poverty (Mehta, 2011). Inclusion of political science as a formal course in preuniversity education and their engaged involvement with political parties over the next decades will inject new life and prospects into these political representations, reinforcing the development of plural and solidary democratic processes. The fusion of democratic principles and social and natural contracts will recover the state's hegemony over the dynamics of predatory economic and political processes, on a local and global scale.

3.4 Environmental Utopia: New Global Nature Contract and Amazonia

The environmental issue reveals the paradigm of ecology as a process of production, development, and reproduction of life. At this point, the environmental utopia is based principally on the confrontation "nature×culture." This utopia presupposes the conception present in the eighteenth-century Enlightenment thought, which virtually takes care to make eternal and invincible "man-nature-culture" in a world marked by great social inequality and an exacerbated consumer market. Rupture of the planet's climate, chemical, and thermodynamic stabilities constitutes a real risk to mankind's future. Mitigating the greenhouse effect and climate change is the principal ecological challenge of this century (Freitas and Silva Freitas, 2013b). The predatory capitalist system constitutes the largest structural contradiction of modernity. This system of capital accumulation, reproduction, and circulation does not have the theoretical and empirical power to safeguard the planet's environmental preservation, today or in the future. The possibility of a total loss of control over mankind's future due to environmental degradation has inspired the introduction of worldwide programs for the preservation and conservation of natural resources, including soils, waters, and the terrestrial atmosphere (IPCC, 2013). This requires developing a new symbolic and material base for this utopia, a base rooted in the principles of responsibility and precaution, globally integrated through mobile structures. A worldwide nature contract will produce new forms of urban organization, novel industrial matrices, professions, and societies. Sustainable development is clearly needed, and Amazonia has a prominent role to play in this contract.

The growing international political tension caused by heavy deforestation and threat of the destruction of Amazonia has induced the implementation of this natural contract. International concern over Amazonia can be expressed through the following statements: Madeleine Albright, US Secretary of State (1997–2001): "When the environment is in danger, there are no borders"; former French President François Mitterrand: "Some countries should give up their sovereignty in favor of the global interests"; former Russian President Mikhail Gorbachev: "Brazil should delegate part of its rights on Amazonia to the international organizations"; and former US Vice President Al Gore—Nobel Peace Prize: "Contrary to what the Brazilians think, Amazonia is not only theirs, it belongs to all of us."

Recently, German Chancellor Angela Merkel and French President Emmanuel Macron also expressed their political concerns about Amazonia's continued destruction. It follows that the deterioration of the environment in Amazonia strengthens the case for its definitive internationalization and insertion into a new worldwide nature contract. The political incompetence of Brazil's new president to develop Amazonia without destroying it raises more tensions in this international framework unfavorable to Brazil. It seems that Brazilian Amazonia's future is a global tragedy foretold.

3.5 Social Utopia: New Worldwide Social Contract

Developing multicultural programs and combatting the growing social inequality are the main linchpins of social utopia. The disintegration of the world`s social base underlines the need to create the operational mechanisms for this "redemptive utopia" whose focus is to value social processes. The tendency of markets and nation states to create new instruments to increase social inequality, incorporating it into public policies, has to be reversed. This enterprise requires synchronizing the marches of state, market, and society from the society's perspective (Vivien, 2001).

Intensification of the migratory movements and growth of violence as instruments of ideological domination and the social exclusion process are issues that, certainly, will figure as part of this century's socioeconomic scenarios. The constant presence worldwide of fragments of social and ecological patterns, in literature, in the arts, in the media, and in the information networks, will be a mark of this period in universal history. Reproduction, food, health, water, energy, education, labor, and ecology are inseparable elements of mankind's future and will continue to be part of governmental guidelines. However, adequate articulation is projected for the economic sciences with these human dimensions valuing the social context (Morin, 2000).

The eco-arts are very important for a new social contract. Communicating science; creating empathy for the natural environment; motivating people to live more sustainably; enhancing values, beliefs, and attitudes; changing habits; constructing self-identities; generating community involvement; and providing alternative modes of consumption (EcoArts-Australis, 2019) are the challenges that face the eco-arts over the next decade.

The disarticulation of education, as a reproductive agent of an alienated and predatory capitalist society, having the categories of violence and poverty as heuristic elements (Marmoz, 1984), will also be present in all complex social studies.

A worldwide social contract should prioritize the fight against the eight "plagues" of postmodernity: racism, poverty, war, structural unemployment, ecological destruction, child labor, endemic and epidemic diseases especially AIDS, and the moral crisis. Global integration of basic public policies—health, education, and culture—and mobilization of world public opinion constitute the main challenge to the operationalization of this utopia. Increasing migratory movements, recurrent violence, and world economic asymmetry underscore the importance of this utopia. They also point to the urgent need to transform the foundations of education as agent of the processes of production and reproduction of predatory capitalist society. The intertwining of social and nature contracts will mitigate worldwide social tensions aggravated in the twenty-first century.

3.6 Ethical Utopia: A Pluricultural Ethical Contract

Self-determination, political and religious tolerance, and respect for differences and for human dignity are the foundations of the utopia of ethics. Its operational mechanisms are rooted in the collective responsibility for constructing a generous, fraternal, solidary, and perennial world, from the multiple features and multicultural coexistences. Its material base presupposes a new market regulation and a restructuring of the state's role in implementing and managing public policies. Its symbolic representation has the arts and education as its main references. The sustainability principle has as its central presupposition the building of an "ethical modernity" that will end destruction of self-affirmation and the process of guaranteeing mankind's permanence on earth, in the sense that ethical modernity is not simply technical modernity. This ethics can be construed as a radical criticism of the notion of destiny, intertwining intelligence and freedom in a virtuous link with goodness. However, universal ethics raises major issues vis-à-vis the globalization of technical and scientific culture. Methodologically, modern science presupposes the distinction between fact and value, physical reality and value, maintaining a strictly extrinsic relation with the sphere of goodness (Ospina, 2000). A new society more involved with mankind's destiny constitutes an immediate application of this utopia. This will mean a rupture with the moral foundations of Western culture's alienating capitalist tradition.

3.7 The Future: The Sustainability of Utopias

Thus, the consolidation of sustainability requires a worldwide political contract, the crystallization of participatory democracy in a universal political system, a new nature contract, an adequate social contract, and an ethical contract among the diverse peoples of the world. These are paradigmatic references for the twenty-first century. These contracts will redefine the concepts of citizenship and economic development in this century (Silva Freitas et al., 2015). The sustainability of utopias requires building an ecumenical consensus that appeals to people's hearts and minds, and is present on all institutional agendas, national projects, international forums, multilateral institutions, transnational agreements, and sects and religions, a philosophical and social conception centered on worldwide multicultural and solidary coexistence (Marti, 2017). A principle of complementarity which includes all exclusions, mutual or not, and reformulates the disjunctive and restrictive processes having as reference the dialectic character of human nature and the complexity of social and economic processes. From these references, the sustainability of people, on a physical, psychic, and spiritual level, as well as the sustainability of places, cities, regions, nations, continents, planets and the cosmos (Schellnhuber et al., 2011), and also the sustainability of utopias, would no longer be a utopia;

and why not say it, no longer an illusion. Science, religion, and humanist education have a key role in the reinvention of these utopias.

The relations between education, science, religion, Amazonia, and sustainability will be analyzed in the next chapter.

References

Besley, T., and Persson, T. (2019). 'Democratic values and institutions', *American Economic Review: Insights*, 1 (1), 59–76.

EcoArts-Australis (2019). 'Why use the arts for environmental sustainability?', <http://www.ecoartsaustralis.org.au/ecoarts-australis/why-use-the-arts-for-environmental-sustainability>, accessed 26 August 2019.

Freitas, M., and Silva Freitas, M. C. (2013a). *Sustainability; Man-Amazonia-World*. Baltimore: America Star Books, Baltimore.

Freitas, M., and Silva Freitas, M. C. (2013b). 'Sustainability, Amazonia and environment: propositions and challenges', *International Journal of Environmental Studies*, 70 (4), 467–76.

Helleiner, E. (2011). 'The limits of incrementalism: the G20, the FSB, and the international regulatory agenda', *Journal of Globalization and Development*, 2 (2), 1–19.

IPCC-Intergovernamental Panel on Climate Change (2013). *'Climate change 2013 – the physical science basic report', Working Group I, 23 to 26 September*, Stockholm, Sweden.

Lin, J. (2013). 'Industrial Policy Revisited: A New Structural Economics Perspective', *Revue d'économie du développement*, *21*, 55–78.

Marmoz, L. (1984). 'L'inefficacité croissante de l'énseignement en France (The increasing inefficiency of education in France)'. Ph.D. thesis in Humanities presented at the University of Caen, France.

Martí, G. (2017). 'New concepts for new dynamics: generating theory for the study of religious innovation and social change', *Journal for the Scientific Study of Religion*, 56 (1), 6–18.

Mattelart, A. (1996). *La mondalisation de la communication*, p. 112. Paris: Press Universitaires de France.

Mehta, P. B. (2011). 'Do new democracies support democracy?', *Journal of Democracy*, 22 (4), 101–13.

Morin, E. (2000). *Les sept savoirs nécessaires à l'éducation du futur*, p. 272. Paris: Éditions Du Seuil.

Ospina, C. L. (2000). *Reunión Internacional de Expertos en Educación Ambiental – nuevas propuestas para la acción*, pp. 30–1. Santiago de Compostela, Spain: Editions UNESCO.

Schellnhuber, H. J. S., Messner, D., Leggewie, C., Leinfelder, R., Nakicenovic, N., Stefan, R., Schlacke, S., Schmid, J., and Schubert, R. (2011). *World in transition – a social contract for sustainability*. Berlin: German Advisory Council on Global Change.

Silva Freitas, M. C., Freitas, M., and Ruiz, M. A. (2015). 'La politique publique pour le développement durable: l'Amazonie, un cas illustrative', *Innovation: The European Journal of Social Science Research*, 28 (2), 192–210.

Vivien, F.-D. (2001). 'Histoire d'un mot, histoire d'une idée: le développement durable à l'épreuve du temps'. In M. Jollivet (ed.), *Le développement durable, de l'utopie au Concept*, pp. 19–60. Paris: Elsevier.

CHAPTER FOUR

Science, Religion, Amazonia, and Sustainability

4.1 Introduction

The analysis of the relations between science, religion, and the origins of the universe and life in the context of sustainability are complex and controversial. It covers the evolution of the genesis of the natural sciences to the improvement of techniques, societies, and mankind. The history of civilizations is strategic to understanding the articulations between science and religion, emphasizing its contradictions, ruptures, and links with sustainability and Amazonia. Highlight to the structural divergences and convergences between science and religion. The rupture between science and religion is a mankind's broader seminal question. The confrontation "nature×culture" is illustrated through scenarios that demonstrate Amazonia's ecological, theological, and scientific importance to the current civilizational framework driven by sustainability and complex philosophical issues. Finally, the importance of sustainability to education for citizenship, to religion, to mankind, and to the planet's future is presented. Given this complex panel, this study adds new elements to the dialogue between science and religion, and education for sustainability (Freitas and Silva Freitas, 2018).

4.2 Relevant Scenarios for Science Education

Economic development models based on the intensive exploitation of natural resources have outlived their usefulness. The possibility of destruction of the human species because of planetary ecological destabilization has obliged societies and governments to incorporate an environmental dimension into all public policies. A new world cycle of science and technology is emerging centered on sustainable practices. The theoretical and empirical elements from the natural sciences that contributed to this new cycle can be pinpointed along the axes described below.

The mathematical physical theory built by Max Planck to reveal the processes related to the phenomena of transport of luminous radiation led Einstein to postulate that electromagnetic radiation was formed by particles without mass and without electric charge, denominated photons. Contrary to the foundations of mechanics, reference at that time, this conception establishes that the correct reading of natural phenomena has to be done in scales, that is, the same phenomenon can be shown in different forms according to the scenarios and the intensities of the constituent elements of the process under analysis. In his studies on radiation, Planck reintroduces, in creative and compelling form, the category "discontinuity of matter" as revealing mathematical structures designated "symmetries," revisiting the Platonic conception on solid and consistent theoretical bases. This study introduced new perspectives to the understanding of the properties of matter properties. The sophisticated atomic model proposed by Rutherford, the results foreseen by Planck's theory and Bohr's theory, that emphasized the discreet character of electronic orbits and other fundamental physical entities to understand the microscopic properties of matter, led to more advanced and sophisticated theoretical and empirical studies on the atomic scale. Heisenberg, Schrödinger, De Broglie, Compton, Born, Gibbs, Dirac, Pauli, Landau, Fermi, and others constructed the following steps, not necessarily in this order.

The first among several contributions established the limitations of classical theory applied to the physical and chemical phenomena known at that time (Heisenberg, 1987). Through the physical relations denominated "uncertainty principles," it was possible to establish when physical theories could be applied to the processes on an atomic scale or to the world accessible to our senses. Using the foundations of a noncommutative algebra and representing the physical entities by womb structures, Heisenberg substituted the classical concept of orbit for one of "quantum state," associating it with experimental measures. Unlike Heisenberg, Schrödinger, using the self-function and self-value representations used by mathematicians and physicists during the eighteenth and the nineteenth centuries, developed a propositive and explanatory mathematical theory for description of "electron waves" (Heisenberg, 2000; Schrödinger, 1990). These entities are fundamental to the study and understanding of the physical and chemical properties of matter and the construction of new electronic devices. De Broglie's contribution is especially interesting: "in same form that to the existence of any particle is associated a wave, every wave state is associated to the existence of a particle." Other scientists assumed the difficult task of building the solid basis of the field of knowledge that later on would become known as quantum mechanics and applied to valid problems on the atomic scale. The introduction of the probability concept in the descriptive language of quantum mechanics, has sophisticated the reading of the current effects of natural phenomena as they now have probabilities of occurrence, and therefore of experimental verification, eliminating definitively the causal character and the deterministic nature, at least on the atomic scale. It is also intriguing to understand the conclusions of the research developed by Compton that confirmed the dual behavior of matter. In appropriate conditions, matter can be presented in oscillatory or corpuscular form, depending on the scales involving the dynamics of the physical process in subject. Nature's several physical properties and characteristics were revealed. The foundations that guide the interaction of light, electromagnetic radiation, with matter, the conduction of electricity and heat, elasticity, magnetism, and other known aspects of the atomic/molecular structure of matter were explained, creating possibilities for technological innovation, particularly for the electric-electronics industry. This industry would become the corner stone of globalization processes. The semiconductor industry has expanded on an exponential scale. Advances in optic lithography have made possible the manufacture of transistorized circuits with greater performance and facilitated the miniaturization of electronic devices. The amplification of the power of microprocessors has greatly impacted technological sectors, especially those related to acquisition, storage, processing, and transmission of information. The projection of a promising future for microelectronics had a great impact on the technology and on world economy. Another fundamental contribution is Einstein's theory, in which the apparent non-joining of space and time concepts, which is generally admitted, is due to very specific and known conditions on space and temporary scales that we are subjected to in our daily lives. These two concepts, which are articulated to each other by the means of light speed, constitute a space-time structure. This structure, in the presence of matter, provokes the curvature of space demanding the use and the incorporation of a new geometric language, a different Euclidean metric to the physical reading of nature's fundamental laws (Born, 1964; Persson, 1996). Through the theory of relativity, Einstein also showed that energy can be transformed into matter and vice versa, indicating that energy is the "universe's fundamental unit."

Finally, the greatest conquest of cosmology to date: "The Big Bang" Theory or "The Great Explosion" proposed by Aleksander Alexandrovich Friedmann and Georges Édouard Lemaître, in the 1920s (Buchwald, 2017). This theory posits the origin, the beginning of universe, as a great explosion, which occurred about 13.7 billion years ago. Ever since then, the universe has been continuously expanding in all directions, with its average temperature decreasing continually. In spite of the controversies surrounding this theory, it has been strengthened by countless discoveries and astronomical observations. The possibility of "the whole" emerging from "nothing" and the unmasking of the dynamics of the creation of the universe, close to singularity, with the projection of real and virtual scenarios of the whole system starting from a part of it, are emergent issues of great significance to current studies on cosmology and on the history of life (Silk, 2000 ; Krauss, 2012; Kêhiri and Pãrõkumu, 1980).

A paradigmatic case is that of the development of biology. The end of spontaneous generation by the Louis Pasteur; Darwin's idea of unifying the natural history of life (Darwin, 2009); and the emergence of heredity in the DNA physical, chemical, and biological structures as proposed by Watson and Crick (1953) have, definitively, given new direction to the history of universal thought.

From these landmarks, the natural sciences have established an incessant and fertile dialogue with all the other scientific fields. It has also created contentious and lasting articulations with philosophy, ethics, religion, and the economic and political processes.

The integration of environmental issues into the world economy and the control of the planet's ecological future are questions that have exerted pressure on current thought systems and started to gain more transparent contours in this century. The technical aspects that have contributed to sustainability have become a field of knowledge essential to the understanding of postmodernity's complex issues. The politicization of sustainability involves not only questions that have been articulated since the foundation of Western and Eastern cultures, but also the contradictions involved in current religious confrontations.

In spite of nature protection policies, the worldwide process of biodiversity deterioration has gained momentum. The main causes are the use of fossil fuel; inadequate use of soils and waters; commercial overexploitation of some species; introduction of predatory species in nonnative ecosystems; increasing pollution of land, waters, and atmosphere; intensification of agriculture with predatory techniques; reordering of territories; and global climate change.

New religious, technical, and political foundations are needed to put pressure on the economic structures that underpin the increasing privatization of the means of production, and the dynamics of capitalist exploitation. This "private wave" clearly contradicts the notion of management of the planet's wealth over a long period. As it becomes more serious, the accelerated spread of extreme poverty in peripheral countries contributes to the increase in the destruction of the most important ecosystems worldwide. Conditions that guarantee sustainability when applied in these countries, in general, strengthen the "biological conception" and/or condemn these populations to perpetual socioeconomic isolation. A new conception of nature is under development: a technified yet humanized nature, bonded to man and mankind's symbolic representations (Martin, 2015).

Religion, politics, and science also have incorporated the environmental issue into their agendas without breaking with the economic processes (Schwadel and Johnson, 2017). New contradictions, articulations, and utopias for the sciences, religions, and sustainable processes are incorporated into this frame, reaffirming the importance of the humid tropics to this world scenario.

These are uncertain times that require alignment between natural science and religion in order to seek solutions to mankind's complex issues. Benz (March 2017) declares that the "new physics may still raise provocative issues concerning methodological limits and divine action, but do not answer today's most pertinent questions on the divine." This framework requires the construction of new explanatory and comprehensive languages articulating science and religion. This assumption potentiates the presentation and analysis of approaches and explanations on the origin of life and the universe in its movement toward the new worldwide sustainable perspectives.

4.3 Science, Religion, and Sustainability: Contradictions and Ruptures

The rationalization of nature's processes presents new problems for religion. These problems have reverberated in all of mankind's sociocultural dimensions.

Boderie (2005) postulates the reason as an image or a divine ray that shines, presenting a fundamental concept: the divinization of reason. On the other hand, David Ingram (1989), based on Habermas, speculates that "the transitions of the mythopoetic visions to the cosmological, religious and metaphysical conceptions and also to the modern forms of understanding demonstrate a moral and cognitive decentralization (…). With the transition of the archaic civilizations to the developed occurs a decisive rupture, and narratives are replaced by explanations which can be justified with arguments. Monotheistic religions approach the reality under a single unifying principle, demonstrating a strong impulse towards universalism and logical consistency (…). With the advent of modernity, the highest principles

(God, Being or Nature) surrendered to the formal principles of discursive reason (…). The objective world of facts should not be confused with to nature, just as the social world should not be equated to society and to culture."

It is important to identify a new central principle to order man's inclusion in the contemporary world, within a conception that proposes to be sustainable. In this conception, from man's birth to his death, his life and his contradictions as well as his general relationships are developed embedded in their interactions with the natural and social environments (Heisenberg, 1987).

Physical, chemical, and biological processes, in general, are not sustainable. The evolution of nature has a tendency to non-sustainability (Dab, 2007). It presents the impossibility of its self-reproduction in identical replicas of current scenarios. Divine sustainability is opposed to scientific non-sustainability in all historical contexts. Therefore, from a biological perspective, life is not sustainable, it "travels" continually towards to physical death. Human divinization opens up new alternatives to this process.

Man's origin is intertwined with the creation of primordial matter that is imbricated to history of birth and evolution of the cosmos, here typified as the universe. Therefore, to understand the history of matter requires unveiling the history of the universe, conferring historicity on its processes of formation and organization (Freitas et al., 2017).

On the other hand, from a religious perspective, God created the universe, the life in it and, simultaneously, all the laws of nature (McGrath, 2005). God's presence in mankind is very strong, independent of discourse and reason. The philosophy of religion demonstrates the difficulty and the complexity of conceptualizing him. He is presented in different scenarios as a personal being, incorporeal, omnipotent, omniscient, omnipresent, creator and maintainer of universe, infinitely good, eternal, and worthy of adoration (Cassé, 1999).

Hugh Gauch Junior (2009) questions whether science has the heuristic scope to explain the existence of God and the universe. At this juncture, he discusses science's claim to expand and extend the scope of the issues, from ordinary to worldview. There are two broad European traditions about the creation of the universe. The first, derived from Greek culture, asserts the possibility of understanding its formation and order based on logic and human reason. The second tradition asserts that God created the universe, at a definite point in time and ordered according to a series of preestablished divine laws (Davies, 2005; Portugal, 2012).

These traditions, the creation of the universe according to the material and energetic preconditions established in past, and those associated with the actions of God have divergences and convergences in their foundations and their operational mechanisms.

Christian de Duve (2005), 1974 Nobel Prize for Medicine, highlights the thesis of biologist Denton (1988), according to which the universe began in order to shelter life. Other experts attribute the creation of the universe to a random characteristic of nature. English chemist Peter Atkins (1981) states "the universe may have emerged out of nothing without divine intervention. At random." For the English cosmologist Martin Rees (1997), "the universe is unique, cognoscible as adapted to life, in a gigantic set of universes. Produced at random." American physicist Lee Smolin (1977) subscribes to this same conception, but in an evolutionist context. Inevitably, the creation of the universe raises two new questions: "Where do we come from?" and "Where are we going?"

Science has accumulated important though not conclusive explanatory contributions to these issues. Russian scientist Alexander Oparin was the first to explain, in 1924, the formation of the first living organism. In 1953, American chemist Stanley Miller produced in his laboratory a set of experiments demonstrating the possibility to explain the origin of life in terms of the laws of physics and chemistry. Miller showed that it is possible, through organic syntheses and adequate physical conditions, to generate life-producing conditions, without divine intervention. Rolston III (2014) replies that answers to problems of this kind should be scientific.

The possibility of life originating from a single living being, from an initial "protobacterium," or the hypothesis that living germs were sent to earth by other civilizations more advanced than ours, or, finally, the prediction of the existence of physicochemical scenarios less complex but sufficiently sophisticated to enable the formation of the first living being on earth are theses on the history of mankind that are both controversial and reductionist. An astrophysicist's thesis that the universe contains a large number of planets with conditions similar to those on earth and, therefore, with great possibility of the existence of conscious life is still a feasible hypothesis (Jacob, 2002), despite its imponderability.

Evolutionary theory predicts we are descendants of prehuman beings and that life has been present on earth for at least 3.5 million years. The development of life from inorganic matter constitutes a rupture that is still outside the explanatory scope of science. This process, according to the biologists, needs to be complemented with new concepts such as integration, pattern, function and system (Murphy, 2010), and also sustainability.

Earth's position in the solar system and in the universe, its physical dimensions as well as its distance from sun, the characteristics of its atmosphere, its speed of translation and rotation, its material composition, geological diversity, and biogeochemical cycles were determining factors to the flowering and evolution of life on our planet, whatever its origin- divine or biological.

A further contribution of science to the understanding of these questions was the idealization of time. According to Schrödinger (1964), three thinkers stand out in this process: Plato, Kant, and Einstein. Plato was the first philosopher to suggest the idea of timeless existence and to contrast it with reason, as a reality more real than our experience. According to his theory of forms and ideas, our experience is nothing but the shadow of timeless existence. This theory emphasizes total abstraction, pure logical thought, symmetries, and geometries in a timeless mathematical perspective. Kant also idealizes space and time categories. The essential point was to form the idea that the mind or world categories can assume other forms we do not yet understand and which do not imply the space and time categories.

Therefore, to Kant, there are other explanatory appearance orders beyond space and time. In this perspective, according to Schrödinger, the possibility of the occurrence of events underpinned by the processes of cause and effect can be replaced by timeless alternatives, in which the notions of before and after do not make sense.

Einstein showed that energy can be converted into matter and vice versa, indicating energy as a "fundamental physical entity" of the universe, in a cosmological scenario that privileges the space-time structures associated with each other through the speed of light. The continued presence of energy in the material and symbolic structures instigates realities and human divinities and opens up new perspectives for research on sustainability. Schrödinger also raises another question: "What can exist after death?" Experience gives us the conviction that human survival does not resist the destruction of the body, to whose life, as it is known, it is inseparably bound. Therefore, in known experiences- and which necessarily occur in space and time- there is no possibility of anything occurring after death. However, in an order of appearance of which time has no part, the notion of after continues to have no meaning. With this condition, there can be no guarantee that there is something after death, but the obvious obstacles to the conception of this possibility can be removed.

Within this new epistemological framework, the scientific questions allow full sustainability without denying the possibility of divine sustainability. The possibility of mankind's extinction due to the planet's ecological destabilization contradicts its absolute divine existence. Maintaining life's permanence on the planet, in all its dimensions, requires a new scale of needs and values aimed at building a new existential relationship between man and nature. The world requires a new aesthetic conception and civilizing framework centered on structures, systems, and sustainable processes. At this juncture, the notion of sustainability can be applied, ad hoc, to the universe and to earth in their totalities, considering them as reciprocal extensions of human existence committed to our common future and in a collective process of material and spiritual sharing. Conceptions that will generate historical actions converge to construct a "cosmic philosophy" of economic development whose central focus would be the scientific and spiritualized paradigm for sustainability.

This singular aspect refers to the reinvention of the myths of immortality, the paradise-world, the spectacle-universe, and the fashion-world. These elements move the imagery of this new civilizing frame. They have, in order, as central references, religion, Amazonia, science, and the virtual merchandise as matrices of its symbolic meanings. It also has ecology as the principal sign of modernity.

An important part of the sustainability's role refers to the existence of complex cultural and biological diversities worldwide. Building a political and religious unity encompassing these cultural and historical diversities in a sustainable perspective presents a great challenge for mankind. However, the operational conditions required by sustainability consider the category "nature" broader and more sophisticated than the category "environment," which requires new methodologies and representations for the analysis of the relationship of type: man-nature-spirit.

Like Kant, Bohr (1929) believed that life cannot be reduced to mechanical, molecular, or genetic structures. At this juncture, the convergence of new scientific and religious approaches in studies of this nature is also a challenge for the specialists.

William Lane Craig (2009) presents six scenarios in which science and religion are mutually relevant: "Religion provides a conceptual frame in which science can flourish; Science is capable of both challenging and confirming the affirmations of religion; Science finds metaphysical problems that religion can help solve; Religion can help to decide among scientific theories; Religion can expand the explanatory capacity of science; and, Science can establish a premise in an argument that produces a conclusion of religious significance."

Ankur Barua (March 2017) asserts that "The key challenge in the theological investigations into the natural sciences's foundations is to spell out the relations between scientific inquiry and Christian doctrine in such a way that the latter illuminates the former without violating its cognitive autonomy and methodological integrity." Wildman (2014) raises important issues about religious naturalism. These issues reaffirm the inseparability of nature and culture. He puts the importance of naturalism into religion, presents the conceptualization of religious naturalism, and analyzes the impacts of religious naturalism on human symbolic representations, morals, and political life. He also does a survey of its importance for our future. In this projection, he indicates civilization's commitments to planetary ecology, and therefore, in our understanding, to sustainability, in all its natural and cultural dimensions.

In this century, the fictional and mechanical language of science is intertwined with the divine and sensitive representations of religion, to form a new universal framework. There is, therefore, something to be gained by initiating a dialogue on sustainability and religious education, in its structures, language, and praxis. This framework, powered by sustainability, requires the presence of ecology and Amazonia.

4.4 Sustainability: Amazonia, the Myth of Immortality and the Sacred

The human condition is imbricated to the nature. Possessors of the knowledge of good and evil, Adam and Eve, lost the divine secret of eternal life (Biblia Sagrada, 2012*s*). Judeo-Christian culture has a matrix, a world genesis or origin centered on the loss of perfection, the loss of Paradise. In this context, a singular issue refers to the theological significance of Amazonia.

The fictional location of the Tree of Life, immortality, and the Tree of Science, good and evil, in Amazonia reserves a special role for its worldwide symbolic and material representations. Its importance grows as models of economic development based on the depreciative use of nature are in the process of collapsing. In this sense, the world's role in Amazonia is also related to the symbolic links that weld this region to the foundations of Western culture (Freitas, 2010).

Euclides da Cunha (2006), an important Brazilian writer, says that "Amazonia is the page that God did not write in 'Genesis- the origin of life'."

Amazonia's role in world is rooted in its ecology and cultural diversities, which has transformed this region into the world's largest living library. Brazil's role in Amazonia has been reaffirmed as inducer of structures, systems, and processes that fuse Amazonian culture to the new framework's foundations from universal thought. It also reaffirms the possibility of designing and building a project for a new ethical, multicultural, and ecumenical world (Freitas and Silva Freitas, 2016).

Science and technology's depreciative use for the destruction of nature conspires against the permanence of the human species, even though they are indispensable in the building of feasible solutions to this tragic situation. The Tree of Knowledge subsumed the Tree of Life, destabilizing the relationship between man and nature. The myth of the blessed: prodigal nature, health, harmony, unity, and immortality is being deconstructed by the rupture of the axis of communication between the "heavenly" and "earthly" dimensions. Having opted for wisdom and not for immortality, the human condition currently finds itself hostage to the economic systems. It uses scientific and religious processes in a nihilistic and privatist perspective.

In this sense, it is also important to investigate the articulations between religious phenomena and ideological processes (Durkheim, 2008). This perspective when placed in the fields of sociology, anthropology, and the natural sciences applied to different conceptions of sustainability establishes new approaches and forms of intervention in reality.

Judeo-Christian and Islamic traditions have common origins; they share many concepts, values, and teachings. In the conception of these religions, a human being occupies a dimension separate from the rest of creation (Lecteur de Pic, 2005). It has a beginning and an end. It is the world's guardian, including nature. Eastern religions, such as Buddhism and Hinduism, preach peace, defend the principle that the man is a constituent part of nature. These relations between religions, nature, and man support permanence, which also can be translated as its sustainability being centered on man-nature-God-culture.

For religions, nature is a sacred entity that must be protected from extinction. In this civilizing frame, the Amazonian peoples have reaffirmed their intrinsic fusions to nature and their commitments to the permanence of life and the planet. Religions do not regard nature as an obstacle to the improvement of man and societies; quite the contrary, they presuppose the human condition as an extension of it (nature). Creating the nexus and operational mechanisms among the material and symbolic representations and articulating science and religion among themselves and sustainability are challenges for us all. Science, religion, sustainability, and Amazonia are definitely incrusted in mankind's history; past, present, and future.

Holy Father leads the organization of the Special Assembly of the Synod of Bishops, to reflect on the theme: Amazonia; New Paths for the Church and for an Integral Ecology, which took place in October 2019, in Rome. The planet's future and Amazonia, the peoples of Amazonia, their territories, nature and culture, sociocultural diversity and human rights, traditional knowledge and modern science, justice and public policies, and spirituality and wisdom are themes on the agenda of the Synod. This event also reaffirmed the importance of Pan-Amazonian culture and this region to the world, and presented the guidelines for a church with Amazonian identity. It insists on building a world that is less consumer-oriented and more committed to Christian values. In broad lines, this assembly discussed the prophetic religious dimensions articulated to a new relationship between man and nature in a sustainable way (Pope Francis, 2019).

Sustainability does not involve only environmental issues. This study has broadened its meaning and reaffirmed the importance of science, religion, and education to mankind's future. In this complex panel, Freitas and Silva Freitas (2018) offered new elements to the dialogue between science, religion, and the education for sustainability. Contrary to reductionist theses (Kellert, 2007), the critical approaches show that religion, science and technology, philosophy, arts, economics, and politics are not inscribed in our genetic code. They are social constructions directed to the improvement of societies, mankind, and the common good.

At this juncture, sustainability constitutes the main invention of the twentieth and the twenty-first centuries. Its incorporation into public politics is in course, on all the scales. This presupposition justifies a more consistent analysis of its meaning and theoretical and empirical reach.

References

Atkins, P. W. (1981). *The creation*, pp. 118–22. New York: Freeman & Co..

Barua, A. (March 2017). 'Investigating the 'science in eastern religions: a methodological inquiry', *Zygon: Journal of Religion and Science*, 52 (1), 124–45.

Biblia Sagrada (Sacred Bible). (2012). *Gênesis – as origens (Genesis – the origins)*, 198nd edn, pp. 22–3. São Paulo: Editora Ave-Maria.

Boderie, G. F. (2005). 'Encycle des secrets de l'Eternité, 1571. Cité dans Gorris R, L'Encycle des secrets de l'eternite ou les secrets d'Uranie'. In D. Courcelles (ed.), *D'un principe philosophique à un genre littéraire: les secrets*, pp. 315–7. Paris: Honoré Champion Éditor.

Bohr, N. (1929). *Bohr's atomic theory and the description of nature*, p. 68. Cambridge: University Press.

Born, M. (1964). *Nobel lectures, physics—1942–1962*, p. 37. Amsterdam: Elsevier Publishing Company.

Buchwald, J. Z. (2017). *The Oxford handbook of the history of physics*. Oxford: Oxford University Press.

Cassé, M. (1999). 'Le cosmos: conceptions et hypothèses'. In E. Morin (ed.), *Le défi du XXI sièche – relier les connaissances*, pp. 27–32. Paris: Editions du Seuil.

Craig, W. L. (2009). 'What is the relation between science and religion?', <http://www.reasonablefaith.org/what-is-the-relation-between-science-and-religion>, accessed 20 July 2017.

Cunha, E. (2006). *À Margem da História (Out of history)*, p. 34. São Paulo: Editora Martin Claret.

Dab, W. (2007). *Santé et environnement, Collection: Que sais-je?* Paris: Puf.

Darwin, C. (2009). *The origin of species*. Cambridge: Cambridge University Press.

Davies, P. (2005). 'Entrapercevoir l'esprit de Dieu'. In J. Staune (ed.), *Science et quête de sens*. Paris: Presses de la Renaissance.

Denton, M. J. (1988). *Nature's destiny. How the laws of biology reveal purpose in the universe*. New York: The Free Press.

Durkheim, E. (2008). *The elementary forms of religious life*. USA: Oxford University Press.

Freitas, M. (2010). 'Amazonia: the nature of the problems and the problems of nature', *International Review of Sociology, Revue Internationale de Sociologie* 13 (1), 183–98.

Freitas, M., and Silva Freitas, M. C. (2016). *A sustentabilidade como paradigma: cultura, ciência e cidadania (Sustainability as paradigm: culture, science and citizenship)*. Petrópolis: Editora Vozes.

Freitas, M., and Silva Freitas, M. C. (2018). 'Science, religion and Amazonia: education for sustainability', *International Journal of Environmental & Science Education-IJESE*, 13 (3), 275–93.

Freitas, M., Silva Freitas, M. C., Ioris, A., and Castro Júnior, W. E. (2017). *Amazônia (Amazonia)*. Lisboa: Chiado Editora.

Gauch Júnior, H. G. (2009). 'Science, worldviews, and education', *Science & Education*, 18, 667–95. Doi: 10.1007/s11191-006-9059-1.

Heisenberg, W. (1987) *Física e Filosofia (Physics and Philosophy), translated by Jorge Leal Ferreira*. Brasília: Edunb.

Heisenberg, W. (2000). *La nature dans la physique*. Paris: Gallimard.

Ingram, D. (1989). *Habermas and the dialect of reason*, pp. 49–51. Connecticut: Yale University Press.

Jacob, F. (2002). 'Q'est-ce que la vie?' In Y. Michaud (ed.), *Université de tous les savoirs, La Vie*, 4. Paris: Éditions Odile Jacob.

Kêhiri, T., and Pãrõkumu, U. (1980). *Antes o mundo não existia (Before the world did not exist)*, pp. 64–82. São Paulo: Livraria Cultura Editora.

Kellert, S. R. (2007). 'Connecting with creation: the convergence of nature, religion, science and culture', *Journal for the Study of Religion, Nature and Culture* 1 (1), 25–37.

Krauss, L. (2012). *A universe from nothing: why there is something rather than nothing*. New York: The Science Foundation.

Lecteur de Pic, C. A. (2005). 'Secrets de la nature et mystères divins'. In D. Courcelles (ed.), *D'un principe philosophique à un genre littéraire: les secrets*. Paris: Honoré Champion Éditor.

Martin, M. K. (2015). 'The cry of earth and conflict with human cultures: a reflection for Christian religious educators', *Journal of Religious Education*, 63, 25–36 (Published online).

Murphy, N. (2010). 'Divine action, emergency and scientific explanation'. In P. Harrison (ed.), *The Cambridge companion to science and religion*, pp. 244–59. USA: Cambridge University Press.

Persson, L. (1996). 'Pierre Auger—a life to service of science', *Acta Oncologica*, 35 (7), 785–7.

Pope Francis. (2019). 'Amazonia: new paths for the church and for an integral ecology', *News Synod*, 18 May, <http://www.sinodo amazonico.va>, accessed 24 September 2019.

Portugal, A. C. (2012). 'Filosofia da Religião (Philosophy of religion)'. In P. Galvão (ed.), *Filosofia—uma introdução por disciplinas (Philosophy – an introduction for disciplines)*. Lisboa: Edições 70.

Rees, M. (1997). *Before the beginning*. Massachusetts: Perseus Books.

Reid, M. (1995). 'O impacto da conservação da biodiversidade sobre os povos indígenas (The impact of biodiversity conservation on indigenous peoples)'. In A. Lopes, and L. Donisete (Eds.), *A temática indígena na escola (The indigenous thematic in school)*, p. 109. Brasília: MEC/MARI, UNESCO.

Rolston III, H. (2014). 'Creative genesis; escalating naturalism and beyond'. In C. Deane-Drummond, D. Evers, N. Gregersen, and G. Peterson (eds.), *Philosophy, teology and the sciences*, 1(1), 9–35. Tübingen: Mohr Siebeck GmbH & Co. KG.

Schrödinger, E. (1964). *What is life?* London: Macmillan.

Schrödinger, E. (1990). *Problemas da física moderna (Problems of modern physics).* São Paulo: Editora Perspectiva.

Silk, J. (2000). *The big bang*, third edition. New York: Times Books.

Smolin, L. (1977). *The life of the cosmos.* Oxford: Oxford University Press.

Watson, J. D., and Crick, H. C. (1953). 'A structure for deoxyribose nucleic acid', *Nature*, 171, 737–42.

Wildman, W. J. (2014). 'Religious naturalism—what it can be, and what it need not be'. In C. Deane-Drummond, D. Evers, N. Gregersen, and G. Peterson (eds.), *Philosophy, teology and the sciences*, 1 (1), 36–58. Tübingen: Mohr Siebeck GmbH & Co. KG.

CHAPTER FIVE

Nature of Sustainability and Amazonia: Brief Analysis

5.1 Introduction

Exploratory studies on aspects from Western civilization processes, which have contributed to sustainability becoming a fundamental field of knowledge for solution and understanding of postmodernity's complex problems, are presented in this chapter.

It proposes to enlarge the significance of sustainability concept and its application to several scenarios immerged in nature-culture universe transfigured by a development process (Silva Freitas and Freitas, 2015).

Finally, Amazonia and diverse controversies and uncertainties of the twenty-first century concerning ecological issues are analyzed as life's process of production and reproduction.

5.2 Sustainability: Foundations and Principles

Universal history marches through languages and praxis anchored by material and spiritual metamorphoses. The different roles of the categories "praxis, matter and spirit" are effectively dependent on the influence of the dominant conceptions of civilization. They are revealed in a tendency of multiple scenarios that compose the substratum of sustainability. These complex scenarios cover a wide range of subjects and have no apparent connection with each other, such as history and physical reality; symbolism and economy; and science and religion, among others, forming the foundations of sustainability (Vivien, 2001).

Sustainability should be understood as a field of knowledge with potential to comprehend all arrangements and combinations of the flow of symbolic and material elements in a human enterprise. Sustainability presupposes the reality of knowledge and action as entities interwoven and in movement, melding nature to culture, transfiguring the conceptions of the civilizing processes, in quality and intensity.

Sustainability can be divided into two major tendencies deriving from classic and modern sustainability theories. The epistemological difficulties of sustainability theories transcend the elaboration of a theory on environmental preservation, clean development mechanisms, beauty, art, and its multiple cultural articulations. The central issue of sustainability is not restricted to the elaboration of a universal language that facilitates understanding of the relationship between man, nature, and the process of addition of "surplus-value" to a certain type of human labor. Its language should prevent the reduction of this value of human labor to market regulation. It is also needs to stimulate the intercultural nature that permeates the construction of collective projects and man's historical praxis (Bourdieu 2003).

Classic sustainability concentrates on the analysis of the concept "preservation of nature," although there is no consensus on the definition of this concept. Preservation as a transcendental idea, preservation as protection of the planet and mankind, and preservation as an immanent concept constitute examples of the different forms of analysis of classic sustainability (Freitas and Pires, 2012; Silva, 2001).

Sustainability's peculiar aspect is its intercultural nature which configures and, simultaneously is transfigured by the dialectic character of space and time categories, projecting the differing civilizational sustainabilities to be discussed below.

For civilizing processes guided by the traditions associated with a conception that is a prisoner of time and dependent on repetitive actions, on the recurrent nature of events and human acts, there prevails a linear trend. With this mental picture, the nature of time is apprehended in cyclical form, and the future is confused with the past once the self-nature of events obeys this recurrent principle (Arantes, 1981). In this ritualistic universe, arrival is interlinked to departure, tendencies are confused with certainties, and objectivity is chained to determinism.

In contrast, for the philosophical schemas permeated by the prospect of progress, the search for innovation and the construction of what is new, time is allied to space as the foundation of a civilizational process that is committed to the future. Between departure and arrival, there are intermediary stages. In irreversible form, new realities emerge, facilitating the construction of multiple cultural tendencies, which reveal ruptures to the dominant historical order. To be and not to be, life and death, tradition and metamorphoses, [...] local and world are contradictions encrusted on this civilizatory process, which embraces sustainability.

Sustainability's cornerstone is unity. Nature's things are presented to us at the same time as concrete matter. Flowers, for example, have many qualities- color, smell, flavor, forms, etc.- but all these qualities are in a singular physical entity. They are not separated one from the other: smell here, color there; color, smell, and flavor are configured as a unit, even though they appear as distinct. This unity cannot be built mechanically. Each part of a leaf has all the properties of a whole leaf (Hegel, 1983).

Ernest Mach (1838–1916) proposed the elimination of space as an active cause in the frame system of mechanics, the field of knowledge that studies movements. He developed the hypothesis that a material point moves with non-accelerated movement not in relation to space, but in relation to the center of all the universe's masses. There is interdependence among all masses in the universe, as if the "isolated" movement of a body depends on the configuration of all the other bodies in the universe. He reaffirms the thesis of unity and holistic interdependence among the entities that compose nature.

Once they are moved, space and time are apprehended by history, as forms to express relations among events or as substrata of ideas, linguistic structures, myths, people's fiction, human feelings, and relations between communities and people. They can also constitute part of the scientific and technological bases that power the technical world- a universe in which the sciences of nature, especially physics, chemistry, and biology, move in an uninterrupted rhythm. At this point, researchers invent new explanatory grounds for nature's scenarios. These scenarios are being continually created with other senses and ranges (Freitas, 2008), transfiguring sustainability's projections of civilizing processes.

Studies on the qualification of the nature of time are controversial. For Kant (1997), temporal entities are subjective, unlike Newton who conceives time as an objective entity. Newton and Mach had different understandings of mental experiments because of their conceptions of the nature of scientific laws. For Newton, his laws were valid for all objects, at any moment, whatever the configuration and distribution of these objects in the universe. Application of these laws, according to Newton, does not depend on the Universe's possible characteristics relative to its constituents, or on the relations among them. Mach, on the other hand, understands physical laws as having to make correlations among experimental facts. There is no general truth for all objects: these express pure contingencies. Their applications depend on the possible characteristics of the universe.

The mechanical nature of time is a common characteristic of these two conceptions. Kant proposes that the concepts of space and time, in principle, are forms of pure intuition, articulated with our relation with nature and not with nature itself (Kant, 1997). They are indispensable for the description of themselves; and therefore, they presuppose the understanding of our experiences, and are not a result of our experiences. Correlations of time and space scales with constituent languages from sustainability's universe meet while in the construction process.

Edgar Morin built a thought schema that embraces all these human dimensions. This schema can be summarized as a type of universal sustainability that has "man-nature indivisibility" as its presupposition. He also emphasizes a type of educational sustainability for mankind's improvement (Morin, 1990).

Understanding mankind's complexity demands a type of sustainability for mankind in which the representative language of these same processes of nature transgresses economic and political contours. It is committed to the social contexts of places, regions, continents, and the world (Freitas, 2008). In this context, nature and culture are interwoven. During and after their confrontations, their contradictions lose their identities, and they initiate the constitution, in structural form, of a new epistemological universe, creating new sustainability, committed to mankind's future.

Man, time, space, and praxis are interlinked categories, which form the substratum that sustain the historical and scientific realities of our social, economic, and political framework. When these categories embrace ecology they raise new problems for sustainability specialists. Problems of type: What scenarios does sustainability create for people's freedom: scenarios rooted in the power of the individual or scenarios molded by the collective responsibility to build a future for all? In which way does sustainability face the confrontation "totality × unity"? How will space and time configure the new sustainability's scales? How will the sustainability of political asymmetries contribute to the construction and maintenance of economical models? Finally, how to use sustainability to establish the political contours of a new world economic order? These are complex topics that need to be exposed to discussion in order to better understand sustainability.

5.3 Sustainability of Nature: Historical Elements

The relation of man and nature also includes the relation between men in war and peace. Anthropology specifies that the study of the relation between man and his environment can be established using two different methods: the form through which the cultures of interest intervene and are articulated with the habitat or a method that delimits a particular environment, thus analyzing space and time from changes of cultural adaptation within predetermined limits (Meggers, 1987). Anthropology's epistemological approach embraces, necessarily, the interconnection of culture with the environment and the contours of the object of research.

However, the concept of nature is more sophisticated than the concept of environment. This requires using more fields of knowledge and other scientific methods to analyze a relationship of the type "man-nature." The concept of nature demands not only what is external to man; it also demands the articulation of singular and universal searching for a cosmic meaning for human existence.

The definitive separation between nature and man and the subsequent fusion of this duality to Western thought schemes has guided the development of the natural sciences in modern times. This has unified different conceptions of nature in a philosophical schema that admits just one type of nature: uniform and reducible, ultimately, to matter and movement. Descartes, the major reference in this field, disentailed the nexuses between the natural sciences, the human condition and theological precepts, establishing principles and elements that have determined the mechanical characteristic of the natural sciences (Descartes, n.d.). He also conceived the principles that facilitated the reduction of the revelation of nature to the explanatory language of mathematics, thereby creating the technical conditions for the emergence and use of the experimental method as criterion for conferring legitimacy on a scientific theory.

The explanatory success of physical theories developed by Descartes, Kepler, Galileo, and Newton, in confrontation with the religious dogmas that "ruled" the skies, was fundamental to the understanding of the articulations between terrestrial and celestial phenomena. Space and time were assumed as absolute greatness, with the representation of an empty space being substituted by a space with lines, angles, and geometric figures. The incorporation of the concept of distance to the Euclidean foundations of geometry and its association to the movement of real objects with the possibility of experimental legitimacy allowed its application in numerous predictions and problems of practical interest. Problems reverberated in biology, which were promptly revitalized in the Swedish naturalist Lineu's pioneer works (1707–1778); the French naturalist Buffon (1707–1788) and other botanists from the eighteenth century, classifying and naming animal and vegetable species and assigning economic value to nature; the French botanist Lamarck's important studies (1744–1829), author of the first scientific theory of the transformation of species; the scientific revolution associated with the studies of Priestley, Cavendish, and Lavoisier, and other chemists toward the end of the eighteenth century, who developed a better understanding of the respiratory processes of animals and the nutritional processes of plants; the studies on the distribution and dynamics of plants developed by the German physicist and geographer Alexander Von Humboldt (1769–1859) and by Swiss botanist Augustin Pyrame de Candolle (1778–1841) in the early nineteenth century; *The Principles of Geology*, by Scottish geologist Charles Lyell (1797–1875) published in 1832, a book that proposed a more sophisticated version of Lineu's concept of the economy of nature, where he introduced the concept of equilibrium of live species; and the creation of the "law of minimum," precursor of the concept "limiting factor," by German chemist Justus von Liebig (1803–1873) in 1840, the law that establishes that the growth of an organism is limited by the amount of nutritional elements it is supplied with, among other important contributions (Freitas and Silva Freitas, 2013).

Between the seventeenth and the nineteenth centuries, technical conditions were generated that materialized in the course of merging science and economy, having the intensive exploration of nature as its paradigm, in perverse and sterile form to us. Perverse because the human condition was put in a position of political subordination, and sterile because all scenarios projected by the thought schema based on this foundation foresee a future for mankind modeled by the concepts of space, time, and matter.

The consolidation of the theory of electromagnetism and optics during the second half of the nineteenth century and the later development of the theory of relativity demanded the revision of three basic concepts: the inertia system, space, and absolute time. During this same period, there was an important development in the field of biology after the publication in 1859 of Darwin's (1809–1882) *The Origin of Species*. This book presents a theory on the evolution of species through natural selection, and inaugurates a new age of the sciences (Darwin, 2009).

The philosophy of science also shows the importance of organization and delimitation of boundaries of a group of themes touching on "ecology." This category was introduced by German naturalist Ernest Haeckel (1834–1919) in 1869 in his studies on the geographical distribution of species; the introduction of the term "biosphere" by Austrian geologist Édouard Suess in his ecological studies of 1875 and the improvement on this concept by the Russian geologist Vladimir Vernadsky (1863–1945) were also important scientific contributions. These concepts interweave the existence of life and the physical-chemical-biological processes of our own planet. We also highlight the introduction in 1899 by Henry Chandler Cowles of the concept climax, which designates the last stage of the progressive evolution of vegetation, contingent upon this vegetation being free from disruptive or destructive actions.

These scenarios illustrate fragments of the natural sciences in the process of crystallization of ecology on the European continent, followed by a fertile dialogue between the natural sciences and the human sciences. This dialogue has led to the emergence of the sustainability of life.

The sustainability of life reserves a special place for biology organized around the following question: What is life? The possibility that life originated from a single live entity, from an "initial protobacterium," or the hypothesis of live germs being sent to earth by other civilizations more advanced than our own, or, finally, the forecast of the existence of less complex physical-chemical scenarios, but sophisticated enough to make possible the formation of the first live entity on earth- all are controversial theses and very restrictive in mankind's history. Astrophysicists theorize that the universe contains a great number of planets with conditions similar to those of earth and, therefore, the possibility they might have conscious life is still a feasible hypothesis, in spite of its imponderable character.

The dominant concept of the sustainability of life that has been projected is anchored in the deterministic processes of eugenics. These European inventions have as their main basis a concept stemming from the works of Newton and Darwin. For socioeconomic processes, the mistaken transposition of fundamental principles from the works of these scientists, *The Mathematical Principles of Natural Philosophy* and *The Origin of Species*, has contributed to the crystallization of a civilization schema shaped by the principles of exacerbated competition, accumulation, and privatization. These schemata have also contributed to the rapid consolidation and expansion of capitalism during the nineteenth and the twentieth centuries, indelibly tarnishing universal history with racism, poverty, and the ecological destruction of the planet. These complex subjects have forced the articulation between the sciences of nature and global economic and political macro-processes, facilitating the planetary dissemination of models of development created by central countries.

The division of nature into the world of the living and the world of the non-living continues to be an obstacle to the construction of a holistic and systemic theory on the processes of nature. Nature is a process; as each process of choice exhibits a sensitive conscience inside itself, it cannot present an explanation of the character of nature. Nothing can be done but to accept the language that presents it as a speculation (Whitehead, 1998).

The fusion of the human condition to nature's sustainability transforms it into the sustainability of life. Irrevocably, human life incorporated the concept of development into its existential universe.

5.4 Sustainability of Development: New Constituent Elements

Economic processes can be predicted; they have a strong structural dependence on technological innovations. Despite the actions and movements of the political enterprises that give form and content to the development of this predictability, the financial market continues attributing meanings and senses to these processes. In this context, it is possible to conceive of the sustainability of development. This sustainability embraces development as a result of specific historical processes. The act of develop(ment) (move)ment through the acceleration of real and virtual objects in time and space adding "surplus-value" to them does not necessarily result in citizenship, a condition that gains legitimacy only through political action.

The physical base of this civilizational process is developed in the form of waves, with constructive and destructive overlaps of cultural processes. The idea as matrix of "genesis," freedom of thought as agent of concrete conception, and praxis as its operative instrument, mediated by the contradictions embedded in the historicity of human beings, are revealed in the tendencies of real and virtual scenarios of national and global cultures (Hegel, 1983). Scenarios that favor the construction of economic and social commitments constitute the sustainability of development. This sustainability, when stimulated, reproduces itself generating innovations and new contradictions. This process is also apprehended by history, which defines the shape and mediations of realities and virtual aspects of the confrontation "development × culture."

This new tendency of development has incorporated the effects of automation and other types of technological changes in labor demand (Acemoglu and Restrepo, 2019; Atack et al., 2019).

An important principle in the history of mankind foresees that modern societies cannot exist without relations of production. Ideas, customs, and individual qualities of the members of society have been altered and developed under the impact of the relations of production, with the processes that include economic, political, and scientific subjects acting as interdependent axes. Production, reproduction, accumulation, circulation, and relations of circulation are constituent elements of modern theories of development, which have molded the political and economic processes through technology (Sen, 2001). In particular, this is the case of the dominant theories of economic development conceived toward the end of the 1970s, with roots in the process of intensive industrialization, privatization, and exacerbated profit (Szentes, 1978).

A radically different picture is now current. Laymert Garcia dos Santos (2001) affirms that "(…) after the colonizing of nature and the unconscious (…), by means of the appropriation of genetic and digital information, the capital is now colonizing, not the dimension of virtual reality, but the virtual dimension of reality." Technologies of digital convergence constitute a material and symbolic substrate of this new cycle of development.

Ecology, biotechnology, nanotechnology, cybernetics, […] and arts are constituent elements of this new civilizational framework, which also incorporates sustainability in its material and symbolic underpinning. However, through science, poverty and social inequality are being incorporated and pondered under a new "light" culture, a "fashion world" which has been impressed on mankind's thought frame. This archetype has established the predatory market as modernity's main frame of reference. The meanings and senses of the national politics of development are reconstructed, subject to the prospect of the asymmetric growth of the international market.

In this context, the confrontation "development x culture" on a sustainable basis constitutes the major global challenge of modernity. The human condition can be reinvented on a solidary basis. It can also be apprehended as the advance of progress, with the development of elements that the cultural processes have shaped. In other words, sustainable development opposes traditional development that denies full citizenship. It promotes democracy committed to diversity and plurality. It proposes to build a social and economic whole, integrated to the symbolic universe that is central to the human condition (Jimenez, 1977). This form of sustainable development denies the use of nature as a commodity and an eternal source of profit.

Clearly, this form of development is important for Amazonia. The region has been incorporated into the foundations of civilization as a postmodern ecological emblem. As a result of this tendency, speculations have resurfaced questioning Amazonia's role in the world and the world's role in Amazonia, a key region for the future of mankind and for understanding complex global issues involving culture, the natural sciences, and sustainable development.

5.5 Environment, Amazonia, and Sustainability: Guidelines and Propositions

Recent archeological studies (Furquim, 2018) register the occupation of Amazonia since 14,000 BP (BP, Before Present, means before 1950), in addition to the first records of domestication of manioc (*Manihot esculenta*) and peach palm (*Bactris gasipae*). This study records the prolonged period of time the peoples of Amazonia took to adapt to their environments and processes of production. Watling et al. (2018) has reaffirmed the importance of indigenous peoples to the process of building biodiversity and cultural diversity in Amazonia. His studies prove that the occupation and sustainable use of this region is an ancient praxis of its traditional populations. These studies confirm the thesis that Amazonia is a cultural construction of its people with the expected external contributions. Its insertion in world sustainability policies is a continuing process.

Brazil is considered leader in the construction of models of sustainable development. The success of its social inclusion programs, its position as the largest environmental power in the twenty-first century, and Amazonia as the main environmental locus in the world are the principal differentiating factors in this socioeconomic process. Globally, Amazonia comprises the most extensive social diversity and biodiversity found in a continuous area. This represents a third of world's reserves of latifoliaceous forests, a fifth of earth's surface fresh water. Amazonia constitutes an important physical entity for the mechanical, thermodynamic, and chemical stabilities of atmospheric processes on a planetary scale. These are complex problems and still subject to scientific speculation.

Amazonian ecosystems behave like an enormous vacuum cleaner, absorbing, for photosynthetic effect, 250–500 million tons of carbon dioxide gas per year (Gash et al., 1996). Research also indicates that Amazonian forests are the largest worldwide sources of renewable biomass on the earth's surface. Official records show that from 1970 to 2000 deforestation of the Brazilian Amazonia's total area was 16% (80 million hectares), an area larger than France. Deforestation, burning, and the use of soils in tropical areas are problems with global impact. These scenarios demonstrate the importance of fighting against deforestation in Amazonia, reaffirming the necessity of its environmental preservation, and building its sustainable development.

From an ecological perspective, there are three structural problems related to environmental sustainability: the need to transform standard industrial matrices; the planetary extension of these matrices, which inhibits a unified management policy; and the invisibility and intangibility of the dimensions of some ecological problems, such as the destruction of the ozone layer, which are not directly accessible to human senses.

The position of Amazonia in the context of sustainable development can be apprehended in three dimensions. Internationally, Amazonia presents several questions to the world, with emphasis on its status as the largest living natural library on the planet, its socioeconomic representation as surplus-value and symbolic processes, its status as a region strategic to Brazil and the world, its role as a source of recycling and as the planet's thermostat; its physical functioning as the planet's mechanism of climate stability; and the world's major natural laboratory for scientific experiments and sustainable practices.

There are two lines of sustainable development in Amazonia. Consolidation of the first tendency relies on the use of languages, techniques, networks, and technology platforms (mechatronics, robotics, cybernetics, biotechnology, fine chemistry, nanotechnology, photonics, information technology, convergence technologies, and microelectronics) to invent new products and processes. The operational mechanisms of this tendency have as a guideline the fusion of a high technology industrial matrix to the ecosystems of Amazonia, without environmental degradation, humanizing its municipalities, and ensuring that public policies reach its inhabitants. The economic dimension of this tendency is heavily dependent on technical education, science, technology, and innovation.

The second tendency proposes to expand nature's protected areas, under NGO management, ensuring the social and economic inclusion of its inhabitants by means of compensatory social and economic mechanisms, transforming them into forest guards, soldiers of sustainability, and forest gatherers, committed to sustainable practices with no economic or social tradition.

The future of sustainable development in Amazonia depends on its transformation into a Brazilian national project to be developed on three fronts: regional, national, and Pan-American integration; internalization of national institutions in the region; and sustainable economic development. Education, science, and technology are the central axes of this national project. Assumptions of a science policy in the region include integration of Amazonia into Brazil, humanization of man's relation with nature, improvement of the quality of life of the local populations, integration and preservation of Amazonia, solutions to the scientific and technological problems typical of the humid tropics, and development of its economic potential.

Currently, Amazonia is increasingly isolated, impoverished, and abandoned. Its complexity, the high degree of predatory internationalization, and short-term political and economic interests from other regions of Brazil and transnational groups installed in Brazil have prevented the introduction of sustainable development models in the region. Brazil's ineffectual diplomacy and political leaders makes this seemingly endless situation worse.

Amazonia's future is firmly situated in sustainable development with roots in multinaturalism and technology. In the present and the past, worldwide involvement in Amazonia was always based on predatory capitalism. Sustainable development in Amazonia is urgent.

References

Acemoglu, D., and Restrepo, P. (2019). 'Automation and new tasks: how technology displaces and reinstates labor', *Journal of Economic Perspectives*, 33 (2), 3–30.

Arantes, P. E. (1981). *Hegel – A ordem do tempo (Hegel—The order of time)*, translated by R. T. Filho Rubens, pp. 36–42. São Paulo: Editora Polis.

Atack, J., Margo, R. A., and Rhode, P. W. (2019). 'Automation of manufacturing in the late nineteenth century: the hand and machine labor study', *Journal of Economic Perspectives*, 33 (2), 51–70.

Bourdieu, P. (2003). *O Poder Simbólico (The symbolic power)*, pp. 129–41. Rio de Janeiro: Editora Bertrand.

Darwin, C. (2009). *The origin of species*. Cambridge: Cambridge University Press.

Descartes, R. (undate). *Discurso sobre o método (The discourse on the method), translated by Márcio Pugliesi and Norberto P. Lima*. São Paulo: Hemus Editora.

Freitas, M. (2008). 'The deadlocks of Western culture and Amazon Region', *Electronic Journal of Sociology*, Tier Two, 1–17.

Freitas, M., and Pires, J. (2012). 'Formation scientifique des peuples amérindiens d'Amazonie et durabilité: proposition et engagements', *La Recherche en Éducation*, 7, 66–86.

Freitas, M., and Silva Freitas, M. C. (2013). *Sustainability; Man-Amazonia-World*. Baltimore: America Star Books.

Furquim, L. P. (2018). 'Archaeobotany and Social Changes in the Southwestern Amazon during the Middle Holocene'. Master's dissertation, University of São Paulo, São Paulo.

Gash, J., Nobre, C., Roberts, J., and Victoria, R. (eds.) (1996). *Amazonian deforestation and climate*, pp. 549–76. Chichester: John Wiley & Sons.

Hegel, G. W. F. (1983). *Introdução a História da Filosofia (Introduction to the History of Philosophy)*, translated by Euclidy C. Silva. São Paulo: Hemus Editora.

Jimenez, M. (1977). *Para Ler Adorno (To Read Adorno)*. Rio de Janeiro: Livraria Francisco Alves Editora.

Kant, I. (1997). *The critique of pure reason*, translated by Paul Guyer and Allen W. Wood. Cambridge: Cambridge University Press.

Meggers, B. J. (1987). *Amazônia, a ilusão de um paraíso (Amazonia, the illusion of a Paradise)*. São Paulo: Editora da Universidade de São Paulo.

Morin, E. (1990). *Science avec conscience*, pp. 225, 235. Paris: Editions du Seuil.

Santos, L. G. (2001). 'A desordem da nova ordem – aceleração tecnológica e ruptura do referencial (The disorder of the new order—technological acceleration and rupture of reference)'. In: G. Viana, M. Silva, and N. Diniz (eds.), *O Desafio da Sustentabilidade (The Challenge of Sustainability)*. São Paulo: Editora Fundação Perseu Abramo.

Sen, A. (2001). *Éthique et économie*. Paris: Presses Universiaires de France.

Silva, M. C. (2001). *Research project for post-doctorate level submitted to the Capes*. Manaus: Printed text.

Silva Freitas, M. C., Freitas, M., and Ruiz, M. A. (2015). 'La politique publique pour le développement durable: l'Amazonie, un cas illustrative'. *Innovation, the European Journal of Social Science Research*, 28 (2), 192–210.

Szentes, T. (1978). *The political economy of underdevelopment*. Hungary: Akademiai Kiado.

Vivien, F.-D. (2001). 'Histoire d'un mot, histoire d'une idée: le développement durable à l'épreuve du temps'. In M. Jollivet (ed.), *Le développement durable, de l'utopie au Concept*, pp. 19–60. Paris: Elsevier.

Watling, J., Shock, M. P., Mongeló, G. Z., Almeida, F. O., Kater, T., De Oliveira, P. E., and Neves, E. G. (2018). 'Direct archaeological evidence for Southwestern Amazonia as an early plant domestication and food production centre', *Plos One*, 25 July, https://journals.plos.org/plosone/article?id=10.1371/journal.pone.0199868,/ accessed 10 July 2019

Whitehead, A. N. (1998). *Le concept de nature,* translated by J. Douchement. Paris: Libraire Philosophique.

CHAPTER SIX

Predatory Development in Brazilian Amazonia

6.1 Introduction

Amazonia's geohistorical and cultural characteristics are fantastic, although its relations with Brazil are conflictive and of political subalternity. Its complexity and importance to the Brazil's future and the mankind were already presented and analyzed in this book.

This chapter analyzes fragments of Amazonia's socioeconomic dimensions (Freitas and Silva Freitas, 2018). Several phenomena and ongoing processes provide a contemporary configuration of the Brazilian Amazonian region and imprint important innovative characteristics on the region: as space for the appropriation and depreciative use of nature by large projects; as a free trade zone; as an extended territory for geopolitical militarism; as a multiethnic space for diverse nationalities, either partially dependent on or relatively independent of the national state; as ecosystems of planetary importance; as a laboratory and prospection ground for a global scientific agenda; as places for decentralized productive developments; and as areas that are challenged and disputed by agents of global powers. These phenomena and processes reopen issues dealing with the relations between regions and nations. They also reopen geographical and cultural issues in the Amazonian region (Silva Freitas et al., 2015; Freitas and Silva Freitas, 2013).

6.2 Amazonia and Development: Contours and Methodological Approach

Through a prospective study, structural issues indicating the ecological and cultural complexity of Amazonia are analyzed, along with the great regional development projects. It is important to emphasize the bibliographical survey and the description of these development projects, in their initial and more advanced stages, established in Amazonia since the twentieth century; the grouping of the themes, projects, and results of research that has chosen to observe closely Amazonia's socioeconomic development, giving priority to the intellectual contexts from 1980 to 2015; and to the periodization of research efforts that have opened and consolidated fields of knowledge and identification of those that have become reference for the public policies applied in the region; and, finally, the promotion of intellectual contributions in order to carry out comparative studies.

Amazonia cannot be held hostage to its past, marked by bloody colonization and predatory development models. Its disorderly occupation and intensive degradation must cease. Globalization and the great projects have deepened the crisis in the region.

6.3 Predatory Development and the Great Projects in Amazonia: A Perverse Inheritance from the Twentieth Century

Capitalism is in check. New definitions for the concepts of citizenship and economic development are the main challenges for the planet's and mankind's sustainable future. Experts predict a collapse of the planet's climate and thermodynamic stabilities if there are no structural changes in the industrial matrices, productive arrangements articulated with natural resources, and the relation between man and nature. Massive projects established in Amazonia during the twentieth century are paradigmatic examples of what should be avoided as a development model in this century. This condition is a prerequisite for sustainability in this important region.

6.3.1 Predatory Development Projects in Amazonia: Illustrative Elements

The history of Amazonia contains several cycles of organization, production, and economic integration into international capital. These cycles were articulated with the demands and the strategic planning of great transnational conglomerates from central countries. Economic activities in Amazonia occur in a closed circuit, where the traditional and native populations are permanently subjected to extractivist policies, a concentrated socioeconomic structure not integrated to regional potentialities, disorganization of worker representation, and lack of an organized civil society (Grégoire and Monzón, 2017). This has conspired against the consolidation of sustainable development in this continental region.

In general, the massive development projects in Amazonia have generated devastating effects on the living conditions of populations located in their catchment areas. Projects for the construction of highways and railways, mines and poles for mineral exploration, installation of massive hydroelectric plants and the logging industry, great physical and electronic monitoring projects, and unplanned settlement in the region are framed in this perspective. Large-scale profit, predatory regional and national agreements, prevalence of the interests of a corporate segment embedded in government benefits, lack of political transparency, inefficiency of the government regulatory agencies, and lack of public commitment to Brazilian society and Brazil's future are typical characteristics of the great projects that have been established in Amazonia, as demonstrated below.

The processes of possession and occupation in Amazonia by the Europeans during the sixteenth, seventeenth, and the eighteenth centuries are paradigmatic examples. These colonial interventions were characterized by genocides and processes of physical and spiritual expropriation of the peoples and environments of Amazonia. These interventions have not yet been fully explained and repaired (Silva Ribeiro, 2018). It describes the large-scale use and transfer of plants from colonies all over the world to the European imperialist countries during the colonial period (Rangan and Carney, 2012), especially from Amazonia, between the seventeenth and the twentieth centuries.

The construction of the Madeira-Mamoré railway, in the periods 1878–1879 and 1907–1912, resulted in the death of at least 6,500 workers. This railroad, 364 km in length, linking the towns of Porto Velho and Guajará-Mirim and which was ready in 1912, cost the public coffers the equivalent of 28 tons of gold at the exchange rate of the day. This international enterprise constitutes the first major technological project implemented in the "deep Amazonian region." Its objective was to guarantee the trade and export of regional rubber, incorporating Amazonia into the world market.

Foot Hardman's studies (1988) emphasize that "(…) It is well known, (…) the economic importance of Collins's enterprise (North American constructors who contracted the building of the Madeira-Mamoré) in the expansion of American capitalism beyond their national borders. According to The *New York Herald*

(…), for the first time in American history, there is an expedition equipped with American material, financed with our money and directed by patricians, to execute abroad, a public work of great magnitude, (…), the Collins expedition assumed an undeniable role in conjucture after the financial crisis of 1873. The great steel industry recovered its activities, eager for new markets. The US$6 million contract for P.T. Collins with the Madeira-Mamoré Railway acquired, in those conditions, an enormous capitalist conquest (…)."

Ruin and barbarity marked the passage of this bloody phantasmagoria in the Amazonian jungles. This highway with no beginning or end, in a paradise untouched by modernity, had nothing more to transport.

The rubber cycle in the region is from that time. Rubber presented itself as a strategic product for the first industrial revolution consolidated on the European continent from the end of the nineteenth century. During this period, rubber production from the states of Amazonas and Pará reached significant proportions. The production of 394 tons in 1839–1840 expanded to 39,266 tons in 1909, three-fifths of the world rubber production, reaching its maximum of 42,000 tons in 1912 (Rangel, 2000). Public revenues of these two states were revitalized with significant impact on the Brazilian economy itself, at the time. In the first decade of the twentieth century, Belém was the third richest city in Brazil (Jornal do Commércio, 2001).

The subsequent contribution of Asia to the world supply of latex introduced new political-economic ingredients to the worldwide market of this natural product. The invention of substitute synthetic products in 1920 projected a rapid decline in demand and production of this product from Amazonia. Henry Ford's failed enterprise at Fordlândia in the Amazonian jungle marked the end of the age of natural rubber production in the region.

Developed initially in a region on the Tapajós River and later in the municipality of Belterra, both in the state of Pará, from 1934 to 1945, this audacious capitalist adventure in the humid tropics was "swallowed up" by the forest. The planting of 3,200,000 rubber trees in an area of 1 million hectares ceded by the Brazilian government to the international automobile industry did not resist the impacts of contemporary world economic processes. Research and exploitation of precious metals, timber, and soybeans, among others, are also part of this global project, aimed primarily at the development and production of the booming automobile industry in the United States of America. The impact of this cycle on the socioeconomic structure of the States in northern Brazil was significant but incapable to contribute to the creation of a scientific and technological policy that would sustain an efficient and integrated industrial matrix in line with the region's interests and specificities. The economic and political control of this cycle by the

British Empire allowed the appropriation and expropriation of a significant part of the wealth generated by the regional populations.

The following cycle in the region was set in motion by the Superintendency for the Economic Valorization of Amazonia created in 1953 by the Getúlio Vargas government. The concentration of financing the processes of rubber harvesting and marketing did not prevent the growth of jute plantations on the Lower Amazonas in the towns of Santarém and Manacapuru, and black pepper in the Bragantina Zone of the State of Pará. These products were introduced into Amazonia by the Japanese. However, lack of regional infrastructure and a regional scientific and technological policy, in addition to rudimentary extraction techniques, lack of commercial scale production and low market value of raw materials extracted in the region, lack of technological methodologies adapted to the tropics, unhealthy working conditions, lack of technical assistance, and emergence of new materials and substitute products, among other factors, contributed to the failure of this Brazilian government project in Amazonia. The success of the pepper plantations is due more to the Japanese colony's work and the tenacity of the Bragantina Zone of the State of Pará than to governmental support.

New economic scenarios were projected for Amazonia after the transfer of the Brazilian capital from Rio de Janeiro to Brasilia, inaugurated in April 1960, and strategically located on the central plateau of Brazil. The construction of a network of highways through Brasilia linking Brazil's southern and eastern regions to Amazonia dates from this period. The construction in the early 1970s of the Belém-Brasilia highway, which is more than 2,000 kilometers long, definitively integrated Amazonia into the capital's major processes of expansion and circulation. The planned development by the Brazilian military regime to integrate the Amazonian region was also intensified at that point in time. This project used capital intensive partnerships between the federal government and the private sector, requiring massive international financing and large-scale production because of the incorporation of advanced technologies. This type of development was based on the formation of agricultural and metallurgical poles, among which were the exploitations of strategic minerals in the county of Serra do Navio in the State of Amapá, in the Serra dos Carajás mountain range and in the Trombetas River region in the State of Pará (...) and in the town of Pitinga in the State of Amazonas. The instability of oil prices on the financial market, which had impacted the world economy during the 1970s, intensified international pressure, especially from the Japanese industry, for the establishment of an oil complex in Amazonia.

The "Great Carajás" Program, founded in 1980, through Decree-Laws 1,831 and 1,825, demarcated an area of 900,000 km^2 from the States of Maranhão, Pará, and Goiás, for the development of an international conglomerate with the objective of mining and processing nonferrous and ferrous minerals, bauxite, manganese, nickel, cassiterite, gold, copper, and others that form the industrial plant of the iron mine in Carajás, managed by Vale S.A. (formerly Vale do Rio Doce/ Sweet River Valley Company). It is comprised of the mining and metallurgical sector of the Great Carajás Program, Albrás-Alunorte in Barcarena State of Pará, and Alumar, an alumina-aluminum complex in São Luiz, capital of the State of Maranhão. This Program also includes plans for agricultural development through rural enterprises of a commercial and speculative character, cattle ranching and logging, and agroforestry companies that are subsidiaries of steel companies or construction companies that are part of this Program (Hall, 1991). It is the world's largest mineral complex on contiguous land.

The region's great hydroelectric potential, the existence of huge mineral reserves, particularly bauxite, the unconditional support of the Brazilian military dictatorship, and the existence of an uninformed and disorganized civil society contributed to the establishment of this authoritarian project in Amazonia.

Lúcio Flávio (1983) states that "(…) in 1979, the first shipment of bauxite from Rio do Norte Mining to the international market left Trombetas Harbor. It was the beginning of the commercial operation of the first 'great mining project' implemented jointly by the federal government and private initiative in order to accelerate the development of Amazonia." Since then, explorations of gold deposits, iron, tin, cassiterite, titanium, copper, bauxite, uranium, potassium, rare earths, niobium, sulfur, manganese, copper, lead, zinc, and diamonds from Brazilian Amazonia estimated at more than US$30 trillion became part of plan of action of the great global economic conglomerates associated to the metallurgical mining sector. The location in Amazonia of 96% of the world's reserves of niobium, a strategic mineral for state-of-art technological processes, underlines its importance in this sector.

Simultaneously, transportation and logistics infrastructure, ports, and a hydropower network were installed in the region, as essential support for the new surge of development in Amazonia, with special emphasis on the hydroelectric plants: Coaracy Nunes in Amapá, Tucuruí in Pará, and Balbina in Amazonas, all of them controversial and expensive projects at the time.

The mapping of energy potential identified that 85% of electric power generation in Brazil comes from hydroelectric plants, and that the northern region had 50% of this potential. The international context of the oil crisis, in line with the authoritarian developmentalist ideology, prioritized the construction of Tucuruí and Balbina hydroelectric plants as strategic logistics for the Great Projects in Amazonia. "But these two hydroelectric plants are just the first of a hundred dams designed to avail the energy potential of the Amazonian basins. A real tragedy for the indigenous peoples" (Santos and Nacke, 2003). The Eletronorte Company, recently bankrupt and fief of traditional politicians, was created to manage the energy complex to meet the demands of production and services of urban and rural populations of this region of the humid tropics. Its recent privatization conspires against the access of traditional populations to energy distribution services.

Symbol of the military regime and project dating from the early 1970s, the Trans-Amazonian Highway is still unfinished and abandoned, having cost more than US$1.5 billion to Brazilian citizens. More than 2.5 million people, coming from different regions of Brazil, live in the vicinity of this Highway, the majority of them with no access to public policies. According to the original plans, its layout should begin in the city of João Pessoa, State of Paraíba and in the city of Recife, State of Pernambuco, with these two parts joining in the town of Picos, State of Piauí, which would connect with the town of Boqueirão da Esperança, on the border of the State of Acre with Peru, by means of 5,600 km of highway (Villaméa, 2001). Corruption; planning dissociated from regional realities; and interests of regional political oligarchies crystalized in municipal, regional, and national parliaments sustained this national tragedy, conceived in the basements of the military dictatorship. Forty-five years later, this Project resulted in a precarious road "only" 2,500 km in length, connecting two regions of Brazil with the worst social indicators.

In this same period, the Northwest Pole Program was thrust on Amazonia. The Program paved the Cuiabá-Porto Velho highway, pushed through colonization projects addressing land ownership regularization and extended protection and health care to several indigenous populations. This Program, financed by the World Bank, which covered the States of Mato Grosso and Rondônia, impacted several indigenous cultures, accelerating the expropriation of its lands. There was an intense though precarious settlement of migrants from the center-south of Brazil to Amazonia. Around 200,000 people migrated to Rondônia; 80 new cities emerged from this human displacement; logging and mining fronts were settled in this area with deforestation of 2 million hectares in one decade (Mindlin, 1990). Deforestation in Amazonia has already surpassed 75 million hectares with impacts on global climate stability. This deforestation is close to reaching the limit beyond which regions of this tropical forest can undergo irreversible changes in the world's climate (Lovejoy and Nobre, 2018; Sud et al., 1996; McGuffie et al., 1995). Studies show that the major changes in land use and land-cover in Amazonia are of an anthropogenic nature (Klimanova et al., 2017).

The Tucuruí hydroelectric plant was inaugurated in 1984. It produces the fourth largest power in the world, 8,000 megawatts, with a flooded area of 2,430 km^2. It cost the Brazilian government US$5 billion, two thirds of which was financed by international loans. Tucuruí is essential to the operation of the Great Carajás Program, which consists of a network of mining-metallurgical projects operating in the region, as presented in this book. The new geospatial and socio-economic configurations imposed on the region under the influence of this program are very unfavorable to the native peoples, accelerating their extermination and cultural uprooting. Since then, the conditions of the Parakanã and Gavião peoples have deteriorated dramatically (Souza, 2000).

The consolidation of this pole dictated the participation of Amazonia in the world mineral market under unequal conditions. The cost/benefit relation of this cycle could not withstand analysis of its accounts. Lack of political transparency, corruption, and extravagant subsidies involving billions of dollars from Brazil's government have accompanied all stages of this cycle, with immense social and economic damage for national and regional societies.

Lúcio Flávio (1983) estimated that the energy subsidies granted by the Brazilian government to Albrás and Aluminar- two aluminum industries located in Pará which would consume 60% of the electric energy generated by Tucuruí or 3% of total Brazilian consumption [reference to 1983]- could exceed US$1 billion over the next 20 years. This shows the permissiveness of Brazilian governments to internationalist interests at the time.

Similarly, Oliveira (2000) emphasizes that the construction of Balbina hydroelectric plant, located in town of Presidente Figueiredo, 110 km from Manaus, and inaugurated in 1989, "(…) was the most criticized public project recently executed in Amazonia. (…). During its planning period, it was assumed that this plant would supply 80% of the city of Manaus's energy demand between 1985 and 1994, with its 250 megawatt power installations. (…). The average power produced over five years (referring to the period 1989–1993) was 117.5 megawatts, 47% of the production forecast (…). Information released by the Eletronorte Company technicians in 1987 estimated the cost of building this plant at US$839 million, and a year later at US$1 billion. External rates and maintenance costs are not included." The appropriation of indigenous lands as well as the destruction and cultural disintegration of the peoples "Waimiri-Atroari" was one of perverse inheritances of this great enterprise in Amazonia.

The construction of the Belo Monte hydroelectric plant on the Xingu River in Altamira in Pará was estimated at US$3.8 billion. Jirau and Santo Antônio hydroelectric plants, both on the Madeira River in Rondônia, are also being built despite resistance from regional populations. The forecast of the initial operation of these technological enterprises is uncertain, considering the growing pressure from NGOs and organized national and international sectors to stop the work, because of its environmental and cultural impacts.

The Jarí Project is the major symbolic enterprise of this cycle. It was implemented by Daniel Keith Ludwig in 1967 through Jarí Ltda., which in 1970 became "Agri-Livestock and Forestry Jarí," installed in an area of 2,400,000 hectares, at the mouth of the Amazonas River. This Project had as its goal the management and agroforestry exploitation, extraction of cellulose, and exploitation of kaolin and strategic mineral deposits, especially bauxite, on a scale large enough to supply world demand.

The size of the project can be evaluated by its numerical projections. In its first stage, an area of 100,000 hectares was reforested, planting 100 million trees *Gmelina arbórea*. The project envisaged the reforestation of 200,000 hectares, with the second stage being planned for the 1980s. A cellulose factory was planned together with a thermoelectric plant, with the capacity to process 750 tons per day. Rice was also cultivated in an area of 15,000 hectares, with an average production of 4–5 tons per hectare; in addition, there was also a herd of 7,000 head of cattle and 5,000 of buffalo (Garrido Filha, 1980).

Like the other major projects implemented in Amazonia, the Jarí Project left behind a trail of misery that directly affected more than 20,000 people in the region under its influence, with explosive demographic growth followed by a collapse in the incipient local public policies.

Rearrangement and resumption of industry and the world market of cellulose together with the accelerated development of new experimental methods in chemistry, physics, applied sciences, and engineering generated great innovations in the technology of new materials. This impacted the entire primary productive sector, mainly after the 1970s, and led to a worldwide decline in the consumption of noble metals and a slowdown in this new "economic outbreak in the region." Technical difficulties of establishing a large-scale agroforestry program in tropical regions, together with the intense mobilization of organized sectors of civil society and the degeneracy of the military regime imposed on the Brazilian people at the time, contributed to the dismantling of this authoritarian project.

The Manaus Free Zone (today called the Industrial Pole of Manaus) was installed in western Amazonia in 1967. This enterprise was an assembly line and export corridor of the largest transnational electro-electronic groups present in Brazil from the 1970s to 1980s. It mobilized the economic interests of a segment of Brazil's elite segment, and continues contributing to power the economy of the region, especially in State of Amazonas. The concentration of the industrial matrix of this development model in Manaus, its reduced economic and social presence in the municipalities, and the lack of consistent and integrated alternative proposals for the region constitute a political debt of the Brazilian State to the region's population. On the other hand, its clean energy matrix, low environmental impact, and the generation of more than 120,000 direct jobs and 450,000 indirect jobs are aspects that strengthen this development model. It also plays an important part in the maintenance of the forests in the State of Amazonas. This State has more than 95% of its 1,570,745.680 km^2 primary forests preserved.

The generation of US$38 billion in this Industrial Pole in 2013 strengthens its importance, although it is globalized at the expense of fiscal incentives and technological innovations developed in the laboratories of transnational industry headquarters. A proposal to amend the Constitution (PAC) with respect to Tax Reform is at the voting stage in the Brazilian Congress and, if approved, it will have disastrous effects on the economy in Amazonia, especially in the State of Amazonas. As the main industrial matrix and development center in Brazilian Amazonia, the state will have a significant reduction in revenue. The tax incentives of its Industrial Pole, known as the "Manaus Free Trade Zone," will be extinguished and replaced by a tax that will have a major impact on its socioeconomic structure. This PAC, currently in the process of admissibility in the Brazilian congress, will be voted on shortly in special committee (Alfaia, 2019).

The lack of an integrated industrial policy for the region and its strong dependence on tax incentives project a very uncertain and unpredictable future for Amazonia. There is a tendency toward declining social and economic indicators of the region. The unfolding of this picture in Amazonia's ecological preservation still needs to be measured. How this picture of the ecological preservation of Amazonia will evolve still needs careful attention.

Since the 1980s, Petrobras has intensified its research and exploration of fossil energy, in particular oil and gas exploration, in the Middle and Upper regions of the Amazonas River. From 2010, this company has an oil and natural gas production unit from Province of Urucu, municipality of Coari, 370 kilometers from Manaus. Average oil production is 56,500 barrels per day, while natural gas production is 9.7 million cubic meters per day (data 2014). This volume of production makes the State of Amazonas the second largest onshore oil producer and third largest producer of natural gas in the country, and the municipality of Coari the largest onshore producer in Brazil.

There are also numerous reports about the negative impacts of the oil industry on the social structure of the town of Coari, where the gas is stored. Its population grew by 7,000 inhabitants, inducing and accelerating the rise of child prostitution, criminal activity, and infectious diseases (Paulo, 2001).

In the 1980s, the federal government implemented the Northern Frontier Project ("Calha Norte") in Amazonia. This Project had a strategic military objective—to secure the borders and a formal excuse for disregarding environmental protection legislation. It covered four basic points: increase military presence on the border, improve bilateral relations, demarcate the borders, and appropriate indigenous policy towards areas of national boundaries between Latin American countries and Suriname (Oliveira, nd). It was a reaffirmation of militarism that depended on a redefinition of the role of Brazil's armed forces in Amazonia. The counterpoint of the national media and public opinion, to the reaction of indigenous leaders, the difficulties in its operationalization, the sporadic nature of its financing, and, finally, its interventionist conception all conspired against the success of the project which, however, was resumed in 2000. Partnership between security forces and Amazonian universities opened the way to improving its infrastructure and logistics of attending to the basic public policies in numerous isolated Amazonian communities, although this was not its central focus.

Equally important, since the 1980s, Amazonia was incorporated into the globalized green market. It highlights the demands related to tourism, bioeconomics, bioindustry, environmental services, clean development mechanisms, tropical technologies directed to atmospheric processes, climate change, monitoring and environmental management, dynamics of biogeochemical cycles and the heat, and hydrological cycles with impacts on several productive sectors and global sociabilities. The establishment of this economic cycle in Amazonia allowed the construction of the economic tendencies of the cost of the environmental services it provides to the world. These costs, which amount to several hundred billion dollars per year, reaffirm its strategic importance for global economic processes (Canalez, 2018).

Agribusiness is another perverse economic cycle for Amazonia. According to Antônio Ióris (2017a), "Since the 1990s, the State apparatus has changed from a position of defender and main financier of agriculture to become manager of complex production networks and coordinator of insertion of national agribusiness into globalized markets. (…). Despite results in terms of economic growth and circulation of capital, agribusiness is also a process of land concentration, marginalization, pollution and proletariarization (…)." Antônio Ióris (2017b) also analyzes its socioecological impacts that are irreversibly damaging to Amazonia and the planet. He demystifies this global redemptive cycle of Amazonia that cannot provide a food security policy to the majority of the surrounding population that has a very low human development index. To make matters worse, the Brazilian state is increasingly dependent on the economic surplus generated by agriculture and has become hostage of the aura of success associated with agribusiness. The Brazilian Gross Domestic Product (GDP) in 2016 was US$1.77 trillion. Agribusiness accounted for 23% of this GDP, contributing US$415.24 billion. This same year, Brazil produced 213.08 million tons of grain in an area of 5.15 million hectares, most in soybean plantations in Amazonia. Agribusiness continues to advance on Amazonian forests, generating massive deforestation, pollution, social poverty, ecological degradation, extermination of indigenous peoples, and irreversible climate change in the region and to the planet.

Another agribusiness victory over the interests of the populations of Amazonia is the recent approval by the Brazilian Ministry of Agriculture of 51 different types of pesticides, 32 of them banned in the European community (Ansa Brasil, 2019). This release of pesticides and the continuous advance of agribusiness over Amazonia increases the threat of irreversible destruction of the region and its cultures. A positive initiative in the interests of Amazonia's sustainability would be to eliminate the perverse subsidies for the whole agribusiness chain and for the intensive cattle ranching in this region. The government of Brazil, again, has become a lethal weapon against Amazonia.

Since 1997, the "Amazonia Surveillance System," a complex surveillance system and electro-electronic network, has been functioning as a support structure to scientific, economic, and political programs in the 5.2 million km^2 of Amazonia (Santos, 2000). It cost approximately US$1.4 billion, and with it the real-time monitoring of Amazonia became feasible. This system has enabled the development of numerous programs in defense of the territory and cultures of Amazonia. It also supports various public policies that benefit regional communities. In the 1970s, the Brazilian military government had already implemented the Project "Radar in Amazonia" (Radambrasil), which aimed to build an inventory of its natural and human resources.

In general, development agencies tend to omit the precarious social indicators resulting from the great projects in Amazonia. The Human Development Index of its municipalities are still among the worst in world. The report presented to Brazilian society by the Parliamentary Inquiry Commission on Illegal Possession of Land in August 2001 highlights the cancellation of 135 fraudulent records of rural properties in Amazonia. According to this Report, documented reports led to the annulment by the Brazilian Land Justice system, just in 2001, of more 37 million hectares of land ownership in States of the Amazonas and Pará, the two largest states in Brazilian Amazonia (A Crítica, 2001). Estimates made in 2001 projected the possible fraudulent possession of 100 million hectares in Amazonia, one-fifth of its total area, and just under one-eighth of Brazil's territory. A network of corruption and influence trafficking, involving businessmen, sectors of the Brazilian judiciary and political interests has sliced Brazilian Amazonia into great hereditary farms. Most of the land is primary forest. A symptomatic case refers to the arrest of Falb Saraiva de Farias, the largest illegal landowner in Brazil. He owns farms in Amazonia whose size is equivalent to 1.5% of the total area of Brazil, equal to the sum of the areas of Portugal and Switzerland. There is no record of anyone else in the world who owns an area of this dimension (Schwartz, 2001).

A new form of reterritorialization emerges under the system of zoning and the principles of environmentalism. By transcending the public and private domains of property, forces of ecological preservation and protection, legitimized by scientific knowledge, are imposed on the traditional land structure. This is one of the tangible results of a global struggle for the ecodevelopment of Amazonia (Ab'Saber, 1996). The age of sustainability reaffirms this tendency based on the inseparability of man, nature, and culture.

In general, indicators of the social and human conditions of populations in Amazonia are disturbing in terms of income, education, child protection, habitability conditions, access to basic health, and perspectives of occupation of workforce, among others. It can be said that Amazonia's social situation has deteriorated over the last 30 years to the point that it is similar to the poverty indexes of the populations of Central Africa.

Despite the differential health care, the infant mortality rate of Amazonia's indigenous population is 67.7 per thousand live births in 2010. In some regions, this index rises to a rate of 100 deaths, as in the case of the Yanomami from the State of Roraima, the Kayapó from the State of Mato Grosso, and of ethnic groups from the Upper Juruá and the Mid Rio Negro (Barros, 2010). These social indicators from Brazilian Amazonia are alarming, especially when compared with similar indicators from Finland, Iceland, Japan, and Norway that present three deaths, in that same year. The impact of the great projects on populations in Amazonia is dramatic.

The failure of the great development projects in Amazonia as well as the demands of sustainability point to the need for new economic structures in the region. These structures require the vocational and contextualized application of industrial, fiscal, and tributary policies to the national and international markets. They also require infrastructure, logistics, well-established innovation and entrepreneurship programs, efficient productivity, decentralization, and diversification of productive arrangements (Lin, 2013).

For Amazonia, it is fundamental to incubate vocational industries in networks, and to induce support for the strategic technological platforms through public financing. The region's continental and ecological dimensions also require implementing clean development mechanisms, information and communication technologies, and bioeconomics programs. This reaffirms the importance of its integration through a development policy with complex structures, concessions, and privatizations with different regulations and public and private participation in the region. With appropriate and contextualized regulations, sustainable development in Amazonia could incorporate the positive principles and guidelines of China's industrial policy (Lin, 2012).

Integration and economic development with the preservation of Pan-American Amazonia's biomes are urgent. The importance of the cultures of its 250 indigenous peoples, its complex biodiversity, and strategic importance for mankind reaffirm its status as the principal global ecological symbol. The region's future is also heavily dependent on the development of other countries that compose Amazonia. Colombia, Peru, and Venezuela are Amazonian countries with many social and economic conflicts on their borders with Brazilian Amazonia, where illicit practices are rampant, such as the increasing timber contraband as well as the smuggling of other products collected from the forest and its rivers. The trafficking of drugs, weapons, wild animals, and precious minerals and guerrilla movements are also problems for this vast region. The absence of public policies and development projects on these borders contributes to this situation, which generates profound socioeconomic inequalities.

It follows that the development of a sustainable economy in Amazonia depends on structural changes. Sustainable development itself still represents a complex challenge for modern organized knowledge.

References

A Crítica. (2001). 'Grilagem de Terras na Amazônia:</SPAN> <SPAN class=titulo_materia>CPI apresenta relatório final em Manaus (Illegal possession of land in Amazonia: CPI presents final report in Manaus)', *Journal edited in Manaus*, 31 August, p. A4.

Ab'Saber, A. N. (1996). *Amazônia (Amazonia)*, p. 132. São Paulo: Editora Universidade de São Paulo.

Alfaia, I. (2019) 'PEC da reforma tributária é mortal para Zona Franca de Manaus (PEC of tributary reform is deadly for Free Zone of Manaus)', *BCN – Brasil Norte Comunicação*, 27 May, <https:// bncamazonas.com.br/ poder/ pec-reforma-tributaria-zona-franca>, accessed *18 August 2019*.

Ansa Brasil. (2019). 'Governo Bolsonaro libera 51 agrotóxicos e totaliza 290 no ano –rítmo de liberação é o mais alto já registrado para o período (Bolsonaro government releases 51 pesticides and totals 290 in the tear – Release rate is the highest ever recorder for the period)', *Agronegocios-Globo*, 23 July, <https://epocanegocios.globo.com/ Brasil /noticia/2019/07/governo-bolsonaro-libera-51-agrotoxicos-e-totaliza-290-no-ano>, accessed 20 August 2019.

Barros, M. (2010). 'A complexidade amazônica e o desenvolvimento sustentável (Amazonian complexity and sustainable development)". In M. Freitas (ed.), *Diálogos com a Amazônia (Dialogues with Amazonia)*, pp. 139–58. Manaus: Editora Valer.

Canalez, G. G. (2018). Agrossistemas Amazônicos: o valor das florestas (Amazonian agri-systems: the value of forests), pp. 124–36. Ph.D. thesis, Federal University of Amazonas, Manaus.

Flávio, L. (1983). *Amazônia: o século perdido (Amazonia: the lost century)*. Belém: Jornal Pessoal.

Foot Hardman, F. (1988). *Trem Fantasma – A Modernidade na Selva (Ghost Train—The Modernity in the Jungle)*, pp. 113–14. São Paulo: Companhia das Letras.

Freitas, M., and Silva Freitas, M. C. (2013). 'Sustainability, Amazonia and environment: propositions and challenges', *International Journal of Environmental Studies*, 70 (4), 467–476.

Freitas, M., and Silva Freitas, M. C. (2018). 'Regional development for sustainability in Amazonia: controversies and challenges', *Geography, Environment, Sustainability-GES*, 11 (4), 112–31.

Garrido, F. I. (1980). *O Projeto Jari e os Capitais Estrangeiros na Amazônia (The Jari Project and the Foreign Capitals in Amazonia)*. Petrópolis: Editora Vozes.

Grégoire, E., and Monzón, L. (2017). 'Institutionalising CSR in Colombia's extractive sector: disciplining society, destabilising enforcement?', *Canadian Journal of Development Studies/Revue canadienne d'etudes du developpement*, 38 (2), 253–271.

Hall, A. (1991). *O Programa Grande Carajás: gênese e evolução (The Great Carajás Program: genesis and evolution)*, pp. 38–44. Belém: Editora FASE/NAEA/UFPA.

Ióris, A. (2017a). 'Agronegócio no Brasil e na Amazônia (Agribusiness in Brazil and in Amazonia)'. In M. Freitas (ed.), *Amazônia (Amazonia)*, pp. 239–62. Lisboa: Chiado Editora.

Ióris, A. (2017b). 'Amazônia; desenvolvimento, exploração e pobreza (Amazonia; development, exploitation and poverty)'. In M. Freitas (ed.), *Amazônia (Amazonia)*, pp. 263–304. Lisboa: Chiado Editora.

Jornal do Commércio. (2001). 'Censo 2000 confirmou (Census 2000 has confirmed)', *Journal edited in Manaus*, 10 January, p. 8.

Klimanova, O., Naumov, A., Greenfieldt, Y., Prado, R. B., and Tretyachenko, D. (2017). 'Regional trends of land use and land cover transformation in Brazil in 2001–2012', *Geography, Environment, Sustainability*, 10 (4), 98–116.

Lin, J. (2012). *New Structural Economics: A Framework for Rethinking Development and Policy*. Washington: World Bank.

Lin, J. (2013). 'Industrial policy revisited: a new structural economics perspective', *Revue d'économie du développement*, 21, 55–78.

Lovejoy, T. E., and Nobre, C. (2018). 'Amazon tipping point', *Science Advances*, 4 (2), 1.

McGuffie, K., Henderson-Sellers, A., Zhang, H., Durbidge, T., and Pitman, A. (1995). 'Global climate sensitivity to tropical deforestation', *Global and Planetary Change*, 10, 97–128.

Mindlin, B. (1990). 'Le Projet Polonoroeste et les indiens', *Ethnies*, 5 (11), 87–96.

Oliveira, J. A. (2000). *'Cidades na Selva (Cities in Jungle)'*, pp. 159–69. Manaus: Editora Valer.

Paulo, A. (2001). 'Transporte do gás de Urucu: ONG diz que gasoduto é ameaça (Gas transport of Urucu: NGO says gas pipeline is threat)', *Jornal A Crítica*, 19 September, p. A6.

Rangan, H., and Judith, C. (2012). 'Environmental history of botanical exchanges in the indian ocean world', *Environment and History*, 18, 311–42.

Rangel, A. (2000). A Bacia do Mar Doce (The Basin of Sweet Sea). Coleção Documentos da Amazônia (Collection: Amazonia's documents), pp. 28–9. Manaus: EDUA.

Santos, R. O. (2000). 'Monitoramento da Amazônia (Monitoring of Amazonia)'. *A Crítica, Journal edited in Manaus*, 13 December, p. 8.

Santos, S. C., and Nacke, A. (eds.) (2003). *Hidrelétricas e Povos Indígenas (Hydroelectric plants and indigenous peoples)*, p. 47. Florianópolis: Letras Contemporâneas.

Schwartz, C. (2001). 'O maior latifundiário do Brasil: Preso em Manaus, ele tem registro de propriedade de 1,5% do território nacional (The largest landowner in Brazil: arrested in Manaus, he has property registration of 1.5% of the national territory)', *Revista Veja*, Edition 1696, 18 April, pp. 64–7, Rio de Janeiro.

Silva, R. A. (2018). 'Olhares inexplorados sobre a Amazônia no Ensino de Física (Unexplored looks on Amazonia in science education)', pp. 362–424. Ph.D. thesis, Federal University of Amazonas, Manaus.

Silva Freitas, M. C., Freitas, M., and Ruiz, M. A. (2015). 'La politique publique pour le développement durable: l'Amazonie, un cas illustrative', *Innovation: The European Journal of Social Science Research*, 28 (2), 192–210.

Souza, A. (2000). *'As Políticas Públicas, os Grandes Projetos e as Perspectivas de Desenvolvimento Sustentável na Amazônia (The public politics, the great projects and the perspectives of sustainable development in Amazonian)'*, pp. 34–5. Manaus: Printed notes.

Sud, Y., Walker, G., Kim, H., Linton, G., Sellers, P., and Lau, W. (1996). 'Biogeophysical consequences of the tropical deforestation Scenario: AGCM simulation study', *Journal of Climate*, 9, 3225–47.

Villaméa, L. C. (2001). 'Saga Transamazônica (Transamazonic Saga)', *Amazônia* 21, 2 (15), 34–41.

CHAPTER SEVEN

Politics in the Service of Non-sustainability and Amazonia

7.1 Introduction

Large-scale exploitation of nature's resources by economic groups tends not to be sustainable. However, the solution to this problem requires high investments in technological innovations that generally do not attract the interest of these same economic groups. Similarly, the preservation of large natural biomes requires continuous heavy investments in clean development mechanisms and appropriate technologies. Sustainability of these environments and their complex cultures requires international collaboration and permanent monitoring. The non-commitment of economic groups and national states with the large impacts of its pollution on nature also hinder the planet's sustainability.

Today, the environmental tragedies in Brazil have assumed dramatic proportions. According to Brazilian government authorities, from August 30, 2019, to the end of October, more than 4,000 tons of crude oil have been scattered in the Atlantic Ocean by a ship not yet identified contaminating the beaches of the 12 states of Brazil. More than 500 locations in these states, which have about 4,000 km of beaches, have already been contaminated, including 12 conservation units- delimited sites to protect flora and fauna, and contribute to the preservation of species. Vast numbers of fish, animals and seabirds, beaches, and coastal regions have already been affected by these stains with not yet measurable impacts. The delay and the inefficiency of the Brazilian government in combating this large natural disaster have contributed to its increasing expansion. The damage of this tragedy, which also affects the beaches of eastern Amazonia, and to Brazilian populations and natural coastal environments, has not yet been measured (G1 Globo, 2019).

The great environmental tragedies in the oceans have continental reaches. Its solutions require new forms of collaboration between countries. The United Nations Convention on the Law of the Sea needs to be discussed in other structuring bases more committed to the planet's ecological future. The sustainability of Amazonia and the world also depends on international cooperation.

7.2 Destruction of Amazon Fund: The Diplomatic Crisis between Brazil, Norway, and Germany

The destruction of the Amazon Fund is a tragedy foretold for Amazonia.[7] The Brazilian government is the main protagonist of its irreversible destruction. This tragedy is a crime against mankind. Several threats to Amazonia by Brazil's new president, Jair Bolsonaro, have already been implemented. The assimilation of the Ministry of the Environment by the Ministry of Agriculture, the complete opening of Amazonia to the productive forces represented by large enterprises characterized by exploitation and predatory use of nature's resources, and the announced Brazilian withdrawal from the Paris Agreement- signed in 2015 by 195 countries for the planet's climate stability- are political propositions implemented by Jair Bolsonaro, who in this way is contributing to Amazonia's immediate destruction.

Brazil was reinserted into the Paris Agreement, as recently announced by Jair Bolsonaro in June 28, 2019, during the G20 Osaka Meeting, a forum of 20 countries that together represent most of the world economy. This political still needs to be consummated through the assimilation of its multiple effects on Brazil's public policies. Today there are still insurmountable obstacles to its success. Bolsonaro's parliamentary support base, which includes large-scale farmers, ranchers, timber entrepreneurs, entrepreneurs of polluting industries, and major investors in predatory economy, among others, will place numerous restrictions on this new Agreement, preventing its full implementation in Amazonia.

7. The Amazon Fund is a REDD (Reducing Emissions from Deforestation and Forest Degradation) mechanism created to raise donations for non-reimbursable investments in efforts to prevent, monitor, and combat deforestation. It promotes the preservation and sustainable use in the Brazilian Amazonia.

The transfer of the demarcation of indigenous lands and the traditional populations to the Ministry of Agriculture is another official action with tragic consequences for the indigenous peoples of Amazonia. Amazonia will most certainly become a great commodity for Brazilian agribusiness. Other similar government initiatives are being planned. The deconstruction of the public policy of environmental monitoring of Amazonia done by the National Institute for Space Research (INPE) since the 1980s is another perverse example. The Federal government has already announced its political intentions to privatize these services. The transparency and accuracy of the scientific data collected, analyzed, and transmitted in real time by INPE are counter to the new president's compulsive desire to privatize Amazonia definitively. But Bolsonaro has reaffirmed his intention to destroy INPE's role in the monitoring of Amazonia. The disorganization of Brazilian society and the complicity of the news media in its destruction collaborate to increase this political chaos and the uncertainties surrounding the future of Amazonia.

The year 2019 has provided a framework for Brazilian Programs with foreign collaborations focused on sustainable development in Amazonia. The Brazilian government now threatens to terminate these agreements unilaterally, in order to privilege the predatory productive forces in Amazonia, especially. In a treacherous move, Bolsonaro also presents his wish to compensate with financial resources from foreign programs, the predatory productive arrangements that according to him were damaged by sustainable development projects in the region. The statement by Brazil's current Environment Minister, Ricardo Salles, is symptomatic: "(…) At the press conference held at Ibama headquarters in São Paulo, Salles said that, after an analysis conducted by the Environment Ministry, 'irregularities' and 'inconsistencies' were found in one quarter of the 103 projects supported by the Amazon Fund – about 30 contracts. 'There are problems with 100% of the NGO contracts', said the Environment Minister" (Bragança, 2019).

The Embassies of Norway and Germany, the major partners of the Amazon Fund, denied any irregularity in the application of its financial resources or in the development of the Program. The Office of the Comptroller General (CGU), a federal entity that conducts accounting audits of federal agencies, also denied Ricardo Salles's statements. Environmentalists assert that the new Brazilian government is conducting an ideological crusade against NGOs and other social movements that defend the preservation of the environment. Amazonia's opening to predatory capitalism conspires against the success of the Amazon Fund (Prizibisczki, 2019).

The Amazon Fund is a Brazilian Program with important international collaboration. It was created by presidential decree, through Decree Law N. 6527 in August 1, 2008. Financial resources raised by this Program were to be managed by the National Bank for Social Development of Brazil (BNDES). The members of its Guidance Committee are representatives of governments and society, representing eight federal government agencies, the nine states of Brazilian Legal Amazonia and six representatives of civil society. Its administrative and technical structure makes it possible to finance and implement projects that integrate culture and nature within a perspective of sustainable development for the region. This Fund also functions as a REDD[8] mechanism created to raise donations for non-reimbursable investments in efforts to prevent, monitor, and combat deforestation, as well as to promote preservation and sustainable use in Brazilian Amazonia. Its main objectives are to: reduce deforestation rates in Brazil and contribute to the decrease of greenhouse gas emissions caused by this process and by forest degradation. Between 2008 and 2018, a total of 103 projects received support from a support fund of US$ 724 millions, and an amount disbursed of US$ 460 millions by the Amazon Fund. It is estimated that this Fund has the potential to raise up to US$ 20 billion in grants by 2020 (Amazon Fund, 2019). Up to 2017, the Amazon Fund had already received donations from foreign governments and companies that totaled US$ 1,217,577,214.17- Norway being the largest donor to the Fund at US$ 1,142,067,325.62. Germany has already contributed US$ 68,143,672.60 (Amazon Fund, Report 2017, 2019). This Fund has contributed to the implementation of important projects for Amazonia's sustainable development. Its destruction constitutes an act of barbarity against the Amazonian peoples and against mankind.

This Fund has brought benefits to many traditional communities in the States of the Amazonas and Pará, the principal states of Brazilian Amazonia. The poor economic indicators of these populations have improved with integrated projects for sustainability financed by the Fund.

Recent studies by WWF-Brazil show that a further 100 Conservation Units in Amazonia are under threat. These Units are associated with the protection of biodiversity, especially water resources, and with the availability of nature's products for the basic needs of traditional populations (Folha-uol, 2019). Today, Amazonia has 316 Federal and State Conservation Units that protect about 1.4 million km^2 of primary forest, about 22,000 rural communities and their cultures.

8. "REDD (Reducing Emissions from Deforestation and Forest Degradation) incentivises a break from historic trends of increasing deforestation rates and greenhouse gases emissions. It is a framework through which developing countries are rewarded financially for any emissions reductions achieved associated with a decrease in the conversion of forests to alternate land uses" (Parker et al., 2009). "Having identified current and/or projected rates of deforestation and forest degradation, a country taking remedial action to effectively reduce those rates will be financially rewarded relative to the extent of their achieved emissions reductions" (Transparency International, 2012).

There is a great gap between the GDP of the States of Amazonia and the poor living conditions of their populations. Pará's GDP in 2015 was R$130.9 billion, about US$43.3 billion. This contrasts with the average household income in the State, which in 2017 was only R$715, about US$238. Its main productive chains are still the mining complex and agribusiness, both of which cause major irreversible environmental impacts. Poverty is growing faster than the destruction and environmental degradation of its forest ecosystems, which have also taken on disastrous proportions.

The growth of poverty in the State of Amazonas occurs at a slower but no less worrying pace. For this same period, the household income of the State was R$828, about US$276, and its GDP was R$86.6 billion, about US$28.8 billion. Unlike Pará, where deforestation has reached approximately 16–20% of its territory, Amazonas has 96–97% of its forests intact, about 1,392,000.00 km^2 of vegetable cover.

Improvement of human development indicators of Amazonian populations depends on several factors with particular emphasis on value adding and diversification of its productive chains, incorporating more innovations and new technologies appropriate to the humid tropics, with greater productivity and competitiveness on sustainable bases.

Through the extinction by presidential decree of two important councils linked to the Amazon Fund, Bolsonaro structurally alters the Fund's managerial function and the prospect of its continuity. On August 10, 2019, the German Minister of the Environment suspended the transfer of R$155 million (US$38.84 million) to the Amazon Fund, announcing that his government's decision was due to the increase in deforestation of the region (O Globo, 2019). Soon after, on August 15, 2019, Norway suspended transfers of R$130 million (US$32.58 million) to the Amazon Fund (Grandelle and Fibe, 2019). Once again, Brazilian authorities become a lethal weapon against Amazonian populations.

7.3 Mining Companies and the Environmental Tragedies of the Towns of Mariana and Brumadinho

In Brazil, large economic groups have a total contempt for environmental preservation and the social protection of their employees. In the name of job generation and wealth, and in collusion with the supervisory agencies, legislation and programs relating to environmental preservation and protection of human rights are systematically ignored by these economic groups. In general, there is a network of political and judicial corruption that supports this crime against society and mankind, with emphasis on environmental destruction and disregard for human rights by large mining companies. The two tragedies that occurred in the towns of Mariana and Brumadinho are paradigmatic.

On November 5, 2015, the "Fundão Dam," under the responsibility of the mining company Samarco, collapsed, releasing 55 million cubic meters of stored mud. Fifteen minutes later this mud destroyed the village of Bento Rodrigues-population 620, located 8 km from the dam. Over the course of the following three weeks, the mud irreversibly contaminated the "Doce River," reaching riverside towns and causing water scarcity, reduction in fishing, commerce, and tourism. In all, 39 municipalities from the States of Minas Gerais and Espírito Santo were affected, involving 1.2 million people directly, and devastating more than 2,000 hectares of land now useless for agricultural purposes (BBC News, 2019). In Mariana, 19 people died and several communities were destroyed. Experts project a period of 50 years for the Doce River's ecology to be restored. Three years after this tragedy, the mining company Samarco, a Brazilian company for the extraction and processing of iron ore, founded in 1977 and managed by the Brazilian Company Vale and the Anglo-Australian BHP Billiton, still refuses to fulfill its social and economic obligations.

But a worse ecological tragedy was yet to come. On January 25, 2019, the dam of the mine at "Feijão Creek," managed by the Vale mining company, located in Brumadinho, in the Belo Horizonte metropolitan region, burst, starting one of Brazil's worst environmental tragedies of the decade. A total volume of 12.7 million cubic meters of mineral tailings destroyed several communities and natural habitats (G1 Minas, 2019), killing 246 people, with a still uncertain number of persons missing. This environmental crime also degraded 112 hectares of native forests, and irreversibly contaminated 305 km of the Paraopeba River. The contamination spread to the São Francisco River, the most important river in the State of Minas Gerais, with vast social, economic, and environmental impacts as yet not fully calculated (Câmara dos Deputados do Brasil, 2019).

On July 9, 2019, the Courts of Justice of the State of Minas Gerais determined the blocking of R$11 billion (about US$3.5 billion) of the Vale Mining Company, responsible for the Brumadinho tragedy. The Court further assessed the extent of the social and environmental damages and forms of this Mining Company to pay its costs (Jornal Estado de Minas Gerais, 2019). Experts maintain that these legal initiatives are still very sympathetic to those responsible for the tragedy.

The existence of more than 300 mining dams in Brazil without appropriate maintenance and inspection continues putting the population at risk and causing serious problems for Brazil's public environmental policies.

Currently, Brazilian environmental issues have assumed dramatic proportions with regional and national impacts, putting at risk agreements and public policies already consolidated at national and international levels. The ecological future of Brazil and Amazonia and the planet need to be better protected.

This tragedy is a crime against mankind. Several threats to the Amazonia by the Brazil's new president, Jair Bolsonaro, have already been implemented. Attempt to incorporate the Brazilian Ministry of the Environment by the Ministry of Agriculture, full opening of Amazonia to the productive forces represented by the large groups of exploitation and predatory use of nature's resources, and the announced Brazilian withdrawal of the Paris Agreement- signed in 2015 by 195 countries for the planet's climate stability- are political propositions implemented by Jair Bolsonaro, who contributes for Amazonia's immediate destruction.

Although Brazil was reinserted into the Paris Climate Agreement in 2019, this political action still needs to be realized through the incorporation of its multiple effects in Brazilian environmental public policy.

Today, there are still insurmountable difficulties for its success. President Bolsonaro's parliamentary support base, including large farmers, ranchers, wood industry, entrepreneurs of polluting industry, and major investors of predatory economy, among others, will place numerous restrictions on this new Brazilian Agreement, impeding its full development in Amazonia.

References

Amazon Fund-Report. (2017, 2019). 'Electronic Site on the Amazon Fund', <https:// www.Amazonfund.org.br>, accessed 10 August 2019.

BBC News. (2019). 'Tragédia de Mariana: Vítimas da lama sofrem com doenças de pele e respiratórias por contaminação por metais pesados e temem nunca ser indenizadas pela Samarco (Mariana's tragedy: Mud victims suffer from skin and respiratory diseases due to heavy metal contamination and fear they will never be compensated by Samarco)', *BBC News*, 10 February, <https://g1.globo.com/mg/minas-gerais/noticia>, accessed 12 March 2019.

Bragança, D. (2019). 'Após declarações de Ricardo Salles sobre BNDS ele afasta Diretora (After declarations by Ricardo Salles on Amazon Fund BNDS he removes Director)', *Oeco*, 19 May, <https://www.oeco.org.br>, accessed 11 June 2019.

Câmara dos Deputados do Brasil. (2019). 'Danos ambientais do desastre em Brumadinho são detalhados em comissão (Environmental damage from the disaster in town of Brumadinho are detailed in committee)', *Camara Legislativa*, 27 February, <https://www2.camara.leg.br/camaranoticias/noticias/meio-ambiente/573031>, accessed 9 March 2019.

Folha-uol. (2019). 'Mais de cem unidades de conservação na Amazônia estão ameaçadas, diz estudo (More than 100 conservation units in Amazonia are threatened, says study)', *Folha News*, 3 July, <https://www1.folha.uol.com.br/ ambiente>, accessed 4 August 2019.

G1 Globo. (2019). 'Manchas de óleo no litoral atingem mais de 500 locais no Nordeste e Sudeste (Crude oil on the coast reach more than 500 locations in the Northeast and Southeast)', *G1 Globo*, 13 November, <https://g1.globo.com/natureza/desastre-ambiental-petroleo-praias/noticia/2019/11/13/manchas-de-oleo-no-litoral-atingem-mais-de-500-locais>, accessed 18 November 2019.

G1 Minas. (2019). 'A relação dos Direitos Humanos com a tragédia de Brumadinho (The relationship of Human Rights with the Brumadinho town's tragedy)', *Jusbrasil*, 26 January, <https://ebradi.jusbrasil.com.br/artigos/671123205>, accessed 22 March 2019.

Grandelle, R., and Fibe, C. (2019). 'Noruega paralisa transferência para o Fundo Amazônia (Norway paralyzed transfers to the Amazon Fund)', *O Globo*, 15 August, p. 1, <https://oglobo.globo.com/sociedade/2019/08/15/582327-noruega-paralisa-repasses-para-protecao>, accessed 23 August 2019.

Jornal do Estado de Minas Gerais. (2019). 'Justiça determina primeira condenação pela tragédia de Brumadinho (Justice determines first condemnation for the tragedy of Brumadinho town)', *O Estado de Minas*, 9 July, p. 1, <https://www.em.com.br/ app/noticia /gerais>, accessed 2 September 2019.

O Globo. (2019). 'Governo Alemão suspende financiamento para a Amazônia. Alemanha suspende financiamento de R$ 155 milhões para projetos de preservação da Amazônia (Germany government suspends funding of R$ 155 million for Amazonia's preservation projects)', *O Globo*, 10 August, p. 1, <https://oglobo.globo.com/sociedade/alemanha-suspende-financiamento-de-155-milhoes-para-projetos-de-preservacao-da-amazonia>, accessed 15 August 2019.

Parker, C., Mitchell, A., Trivedi, M., Mardas, N., and Sosis, K. (2009). *The little REDD book*. Oxford: Global Canopy Programme.

Prizibisczki, C. (2019). 'Embaixada da Alemanha e entidades ambientais contradizem declarações de Salles sobre o Fundo Amazônia (Embassy of Germany and environmental entities contradict the Salles's declarations on Amazon Fund)', *Oeco*, 17 May, < https://www.oeco.org.br>, accessed 21 July 2019.

Transparency International. (2012). *Keeping REDD+ clean. A step-by-step guide to preventing corruption*. Berlin, Germany: Transparency International.

CHAPTER EIGHT

Nuances of Sustainability: Fantastic Visions from Amazonia

8.1 Introduction

From a humanist perspective, the history of capitalist and communist expansion is tragic. Even today, capitalism continues to destroy the environments and the cultures of the colonized peoples. For a long time colonizers have usurped the histories of the colonized. Their wars and unsustainable use of the planet are a growing threat to mankind's future. In these bloodthirsty actions, the use of arms always supersedes diplomacy and sustainability. Under these circumstances, even the use of science and technology becomes a lethal weapon against specific populations and mankind. The Vietnam War is a particular example.

By some estimates, the forest canopy in Vietnam declined by 50% from 1945 to 1980. From 1961 to 1971, during the Vietnam War, U.S. forces sprayed 72 million liters of herbicides, including Agent Orange, on the Vietnamese countryside and dropped roughly 13 million tons of bombs, according to Jakarta-based forestry expert Chris Lang (Yale E360, 2010). Agent Orange is a powerful chemical herbicide and defoliant known for its use by the U.S. military during the Vietnam War to eliminate forest canopy and crops that sustained North Vietnamese and Vietcong troops. The diseases caused by exposure to Agent Orange now include everything from non-Hodgkin lymphoma, prostate cancer, and multiple myeloma to Parkinson's disease and ischemic heart disease. Traces of Agent Orange are still found in food in Vietnam, 30–40 years after the spraying ended (Spurgeon, 2003).

The herbicides that were sprayed on South Vietnam affected 43% of the cultivated area and 44% of its total area. Seventy percent of the south's coconut groves and 60% of its rubber plantations were destroyed, together with 110,000 hectares of forest and 150,000 hectares of mangroves, along with enough crops to feed 2 million people. It has also been claimed that 43% of the south's plantations and orchards were destroyed, and 44% of the forest wealth, about 5 million hectares (Green Left, 1993). As we have already stated, capitalism has no heuristic reach to build peace and at the same time exploit the resources of nature preserving its natural cycles.

Generations of young people continue to be sacrificed in the name of freedom and a free market, which are the privilege and the hostage of a select few. Sustainability inserts new elements into this process. The correlation of this concept with the worldwide expansion of European culture and the colonial processes, during the nineteenth and the twentieth centuries, is analyzed by the authors. Special prominence is given to the ecological issue as an instrument of ideological domination. Worldwide scientific and educational processes, as instruments of political alienation of people from developing countries, are also discussed. Finally, several scientific and ethnological scenarios from Brazilian Amazonia are presented, and the importance of this region to the contemporary world is emphasized.

8.2 Brief Digression on the Concept of Sustainability

The origins of the concept of sustainability are linked to the expansion of European culture to the Asian, African, and American continents. Science, religion, and politics, associated with the economy, have molded the civilizatory processes installed in these regions. These processes created the political conditions for the revitalization, recreation, and financing of new research programs and scientific guidelines in the central countries. New demands for international markets, and new senses and meanings of consumption emerged (Freitas and Silva Freitas, 2016).

These colonization processes were guided by a concept of evolution with roots in the fight to preserve life, repudiating noncompetitive and peaceful coexistence among peoples. These colonial projects introduced a fallacious concept of biological diversity into the analysis of the dynamics of human populations. This concept permeates the behavior of all biological species, as biological systems adapt to the alterations and disturbances in its environments.[9] Pro-slavery activities, social

9. In general, beginning in the 1980s, biodiversity expressed the variety of existing organisms in the world, including their genetic diversity and the groups that form them (Reid, 1995).

disintegration, compulsory settling, and forced labor were commonplace practices in this period of mankind's history. These processes gave sustainability to the worldwide economic processes emanating from the European continent, at that time.

For this framework, up until the first decade of the twentieth century, two major ethnographical theories originated in Europe, even though they prevailed worldwide. In general terms, they were opposites. Their studies on the organization and dynamics of races were in radical opposition. Environment theory, through its more radical representatives Hippolyte Taine and Thomas Buckle, attributed to the influence of the environment all the differences between men and between peoples. Race theory, on the other hand, defended mainly by Gobineau, considered race as the only decisive origin of all mankind's cultural and ethnological differences (Haberlandt, 1929). Other conceptions mixed elements of these theories, in an attempt to reconcile and harmonize their more extreme positions.

This is, approximately, the synthesis-scenario of the ethnographic conceptions that represented and motivated the Westernizing of the planet at that time.

Restriction of access to the natural resources of each people is a constant in the history of civilization (UNESCO, 2005). The use of natural resources has been a constituent and decisive element of developmental models and the forms of social organization of several peoples (Renan, 2005), contributing to the incorporation of nature into their cultures.

The history of capitalism in Western Europe illustrates the dimension of the destruction of natural resources by its economic processes (Vivien, 2001). A specific case is that of the development of the iron industry and the environmental issue. During the sixteenth and seventeenth centuries and in the first years of the eighteenth century, English forests were systematically torn down for use in the treatment of iron and in naval engineering. Indispensable elements of the industrial revolution were unveiled in that country at the time.[10] The situation assumed dramatic contours... to the point that the Irish, unable to produce the quantity of trees needed to produce the bark for the requirements of the leather trade, were forced to buy bark from England and timber from Norway, and to export its leather products unprocessed. This environmental problem accelerated the development of techniques facilitated by the use of coal, available in huge reserves, as a substitute for the expensive, rare wood. Full mastery of steel preparation techniques, allied to

10. During the period of feudalism, Europe was submitted to an intense process of territorial redivision. In those days, land was the most important means of production, a sign of property, wealth, and power in the society. Together with demographic growth, the rhythm of deforestation was intensified and extended, using the available technical resources of that period (Elias, 1994).

the power generated by waterwheels, facilitated the appearance of large-scale metallurgical production. This contributed to the emergence of the age of the steam engine, and closed the curtain on a period driven by practical needs and primitive empiricism (Mantoux, 1927).

Pressed by new demands and emergent needs, man acted on nature, transforming it, humanizing it, and creating the historical conditions needed to consolidate the industrial revolution. The construction of the steam engine by James Watt in the eighteenth century was the main mark of the industrial revolution at that time.[11]

The interlacing of classical mechanics and thermodynamics during the nineteenth century and with electromagnetism at the beginning of the second half of this same century guaranteed the theoretical and empirical conditions to consolidate and multiply the technological processes, which facilitated the production and expansion of markets on a worldwide scale.

This new civilizatory cycle made the scientific methods a definitive part of the history of technology, accelerating the development of the industrial revolution. A social revolution emerged in England, and new economic and political processes guided the globalization of European culture during the following centuries, creating the historical conditions that caused the serious environmental problems of the contemporary world.

The later displacement and transnationalization of great industries did not prevent the globalization of environmental problems; quite the contrary, in a certain sense, it made them worse with the transformation of the world into a gigantic factory, regulated and maintained by deep and growing relations of social inequality (Ianni, 1999).

The definitive establishment of electromagnetic theory by James C. Maxwell[12] during the second half of the nineteenth century created the scientific and technological conditions for the worldwide access to electricity. This increased the scope of capital and incorporated electric currents and magnets among the world's new political conceptions.[13]

11. James Watt, inventor of the steam machine, was born on January 19, 1736, in Greenwood, Scotland.
12. James Clerk Maxwell (1831–1879), Scottish physicist, founded the electromagnetic theory and contributed to the consolidation of kinetic theory of gases, among several other scientific contributions.
13. At the time, only the concepts of kinetic energy, potential energy, and calorific energy were known. The first concept is associated with the movements of objects, the second is the energy due to the position occupied by the object, and calorific energy refers to the processes of heat transport.

At this point, new perspectives for electric-electronic processes would permeate aspects of human sociability in the first few years of the twentieth century (Abrantes, 1998).

The appropriation of scientific knowledge, a collective good, by capitalist permissiveness, without any investment, contributed to the emergence of this new cycle. The relations between capital and science were observed by Karl Marx (1818–1883) who establishes:

> Science costs absolutely nothing to capitalism, which does not prevent its exploitation. Alienating science is incorporated to capital as alienating work. Capitalist appropriation and personal appropriation, be it of science or of material wealth, are tremendously disparate things [...] (Marx, 1984).

The connection of agrarian issues to industrial super-production presents new problems and dilemmas. The mechanization of intensive agriculture, coupled with the developments in climatology, chemistry, and biology applied to agricultural processes, underwrote the tendency of this new cycle.

Continuing, Marx (1985) asserts:

> [...] One part of society demands a tribute from the other for the right to live on the earth, as well as, in general, being implicit in landed properties the proprietors's right to explore the terrestrial body, the earth's bowels, the atmosphere and, thereby, the maintenance and development of life [...].

But, the transition period to the industrial society also had its major share of contradictions. Between 1730 and 1795, Rudé (1995) counted, through announcements in newspapers, 375 mutinies of all types in England, 275 of them due to lack of food. The conditions for the heightening of tensions between capital and labor were created. These conditions have been intensified in current political systems.

* * *

Full control of world expansion by European capitalism also conspired against the development of peripheral countries. Non-alignment and non-institutionalization of the advances in the sciences and technologies in the national projects of these countries constituted an obstacle to raising the quality of living of significant segments of their populations.[14]

14. The scenarios projected for this century point to the strengthening of American economic supremacy, and the intensification of the world's social asymmetry. In this international frame, the African and South America continents are distant from world markets, unable to maintain a sovereign and equitable political dialogue with the developed countries.

The perverse characteristics of economic and social development in these peripheral regions include expulsion of indigenous populations from their lands, accelerated environmental degradation, concentration and appropriation of large areas of productive lands, current social problems from constant regional migrations, destruction of the historical memory of places, villages, pioneer cities, and the process of dehumanizing spaces and programs of territorial occupation. This is the heritage of colonialism.

During a critical process of political contradictions that marked the end of the twentieth century, Robert Castel (1995) affirms:

> (…) It is important to denounce the State's hegemony over civil society, the bureaucratic and ineffective operation of its instruments, the abstraction of social law and its impossibility to raise concrete solidarities, and force the transformations to take into account particularities of situations and mobilization of subjects (…) to the possibility of rehabilitation to build a project, and define a social insertion contract.

The development models of underdeveloped countries prioritize programs of regional exploration and extraction, effectively reproducing the poverty of the extractors, the socioeconomic non-sustainability of extractive processes, the inefficiency and inefficacy of extractive methods, the instability and financial inconsistency of these extractive processes in the context of regional and national economies, the absence of adequate technical support, the complex physical-chemical-biological characteristics of the regions and territories, of climates, biological diversities, fragility of soils, social isolation, and absence of base industries or regional markets which add to market prices, just socially, for extractive products; all of these being results attributable to the disastrous and extemporaneous actions of the governments of these countries.[15]

This virtual face of sustainability has been reenergized by the technological actions of economic models and by hegemonic governments on a global scale. In general, these governments reinforce the binary ideology typical of colonialist processes, of the type: east-west, north-south, poor-rich, developed-underdeveloped, advanced-backward, and responsible-irresponsible. This discriminatory and ethnocentric ideology constitutes an impediment to the sustainable world. It is as if there were two kinds of sustainability: one for poor countries and another for rich countries (Batista et al., 2005).

15. In his book, Homma (2003) presents the general principles that have guided the economic politics of Brazilian Amazonia. The foundations of his analysis can be generalized to a wide majority of countries with extractivist economies.

Prevalence of binary morals in Western culture favored the emergence of globalization processes lacking the cultural and historical dimension. This has given weight to the thesis of a central government that would confer legitimacy on a global regulation, with roots in the global ethos. This thesis would be guided by the principle of the universality of the rights of man and by sustainable development. However, this very thesis does not take into account the economic determinations and the social interests that articulate globalization processes (Wahl, 2001). This does not eliminate the merit of the sustainable development concept as having facilitated the creation of stronger links between the natural and human sciences.

On the other hand, the concept of sustainable development gains historical validity in local experiences, where there are policies for the use of the resources of a territory, designed to involve social configurations, political situations, and possibilities of the application of available technologies (Dubois and Mahieu, 2002). Universalization of local experiences, through projections on a planetary scale, will come to be regulated by a negotiated common objective: the preservation of biodiversity which, for its time, is most closely associated to cultural diversity (Larrère and Larrère, 1997). The existence of objective conditions for its full accomplishment is still open to discussion. The Rights of Peoples and Traditional Communities are also fundamental preconditions for the success of this type of development (Shiraishi Neto, 2007).

In general, the official bureaucratic discourse projects sustainable development in an essentially ecological dimension, isolating it from its social substratum. Factors internal to the development policies prevalent in industrialized countries are opposed to sustainable development (Schellnhuber et al., 2011). The exportation of polluting industries to developing countries, a consequence imposed by inadequate environmental regulation in these countries, and ownership and control of enormous territorial areas in tropical regions constitute important, contradictory examples.

To illustrate the nuances of the concept of sustainability in the contemporary world, there follows a brief presentation on some ethnographical scenarios from Amazonia (Freitas et al., 2003). Important reference is made to the indigenous peoples and to the traditional population of Amazonia, who have expressed a desire for access to public politics, a dream still unattainable for the large majority of them. It also observed the speed with which the indigenous population acquires and manages scientific knowledge for the benefit of their collectives. It has been demonstrated that community members, despite living in paradise, are not born with original sin. Their faiths and myths are historical. The raising of these populations to the condition of protagonists, to the direct benefits of science and technology, constitutes a human perspective present in all local social vindications (Freitas et al., 2016).

As presented previously, Amazonia's geographical and historical magnificence embraces a multiplicity of biological and social diversities, creating new social and economic perspectives for mankind (Freitas, 2014).

8.3 Amazonia and Sustainability: A Perfect Marriage

Amazonia supports the area of greatest social diversity and biodiversity in the world and is home to one-third of earth's rain forests and one-fifth of its surface fresh water.[16] Amazonia also plays an important part in the mechanical, thermodynamic, and chemical stabilities of the world's atmospheric processes.

There is also the question of the theological significance of Amazonia. The Judeo-Christian culture has a matrix, in which the genesis or origin of the earth is centered around the loss of perfection, "Paradise Lost." The fictional location of the tree of life, key to immortality, and the tree of knowledge of good and evil, in Amazonia reserves a special place for worldwide symbolic representations in this region. Its importance grows as models of economic development based on the depreciative use of nature are in collapse.

It is also important to investigate the articulations of religious phenomena and ideological processes (Durkheim, 2003) in the context of sustainability. This approach when analyzed within the fields of sociology, anthropology, and the natural sciences and applied to different visions on sustainability yields novel mechanisms and forms of intervention in reality.

As presented in Chapter 4, the relations between religion and sustainability are paradigmatic. Christianity, Judaism, and Islam have common origins; they share many concepts, values, and teachings. At the conception of these religions, human beings are dissociated from the rest of creation (Lecteur de Pic, 2005). They have a beginning and an end. They are the guardians of the world, including nature. Oriental religions, such as Buddhism and Hinduism, preach peace, defending the principle that man is a constituent part of nature. These relations between religions, nature, and man underscore their permanence, which can also be translated as sustainability centered on nature-God-culture.

In a religious context, nature is a sacred entity that must be protected from extinction. Within this civilizing frame, the peoples of Amazonia have reaffirmed their intrinsic fusion with nature and their commitment to the permanence of life and the planet. Religions do not regard nature as an obstacle to the perfectibility of man; on the contrary, they presuppose the human condition as an extension of

16. The Amazonas River's average water flow in 1 second is enough for the daily provisioning of a city with 294,000 inhabitants.

nature. The creation of a nexus and operational mechanisms between the global material and symbolic representations that articulate sciences and religions among themselves and to sustainability are challenges for us all (Freitas et al., 2016). The existence of political injunctions in this process of global scope presents additional elements for mankind.

Donald Trump's refusal to recognize climate change injects new elements into this picture. Political discussion of the ecological issue holds the capitalist system in check. The rupture with this political regime has gained new facets this century. In November 2019 Trump begins a year-long process to formally exit Paris climate agreement (Holden, 2019). The political consequences of this action by the US president are still unpredictable. Amazonia's worldwide importance will be strengthened.

Since September 2019 Australian bushfires has been intensified with major impacts on the global greenhouse effect. By the end of January 2020, 39 people and more than one million animals had already been dead in these Australian bushfires. In this context, Amazonia's environmental preservation acquires greater global importance.

In this Age of the Environment, certain scenarios have demanded new meanings and a redefinition of science education. In this context, to redefine means the challenge of constructing new approaches, senses, and forms of intervention in the educational sciences, the natural sciences, and technology in public policies and in production processes.

For Marilene C. da Silva (2013), "Amazonia is a great mercantilist unit of science as a productive force." This statement defines the region as a laboratory, an area with multiple research subjects for the development of the natural sciences. The effort to solve it has not ended; nature is reinvented constantly by the demands for knowledge and the operationalization of the production of wealth. Amazonia has always represented an opportunity to test and evaluate the relations of force between culture and nature, and the international division of the world's scientific agenda and the dilettantism of researcher curiosity." Projects, collections, institutions, and emerging organizations need to articulate statements from fields of knowledge with the civilization process of capital (Wagner, 2019). For the history of scientific thought on Amazonia, these processes correspond to relations in the domain of nature for man, in which the action does not exclude composition in valorization manners and in the development of work on an enlarged scale, as discovery and social cooperation.

The confrontation "nature×culture" in Amazonia has been defined as an important political and economic question. The constant tension between region, nation, and world is a fact that has become common place in Amazonia. Sustainable development is the main agent of this historical process (Freitas et al., 2017).

8.4 Fantastic Visions from Amazonia

A deluge had fallen on the earth, an unbearable heatwave has expanded and overflowed in all directions and places on the planet. Seas and oceans, furious, insisted on not returning to their natural beds. The air, unbearable and rotten, burned the lungs of those that breathed it. The market for environmental services was in constant expansion. Cloning commodities were the exchange markets' stars and the great international financial centers. There were clones for all tastes: hard, soft, tasty, hot, scented, fetid, expensive, cheap, pleasant, unpleasant, and even biodegradable.

The artificial water did not want to flow towards processes vital to the maintenance of life. The soils insist on not obeying the stimuli of artificial products. But, imperturbable, night and day continued as in ancient times, with the stars and sun marking the progress of each.

People, communities, and populations had already, with some difficulty and with the help of science and technology, adapted to this new reality. Cheerful and confident in the future and in a more generous state of social promotion, different people dreamed of a healthier and self-sustainable planet for future generations. Summoned by world authorities, they are prepared for a great battle, a war that seems to be the last of our civilization. Confrontations are expected to solve mankind's main problems.

Amazonia had presented the last resistance to be overcome; Southeast Asia and Southeast and Central Africa had already succumbed. An international army had already marched to occupy strategic positions in Amazonia. The forest, fornication center and guardian of sacred firewood, needs to be protected against the indigenous peoples and the traditional populations, invaders without any voucher or right of ownership of these places so important to mankind's survival.

The mode of attack on Amazonia had already been defined: with the help of modern weapons, we will remove the sky from the region; for whatever period necessary, we will make all its waters rotten and useless for life, infecting them with an unbearable pestilential odor, similar to that of the Tietê River;[17] and finally, we will infest these tropical lands with chemical products that transmit the smell of death through the air, much like that of the city of Cubatão.[18]

We expected that without their sky, water, and land, matrix of their divinities, from which emerged part of their codes, myths, and many of the indispensable existential references for the design of their conceptions of the universe, they (the indigenous peoples) would surrender without resistance. And, more importantly,

17. This River crosses the city of São Paulo, Brazil's main industrial center.
18. Industrial center in São Paulo State, considered to be one of most polluted cities in the world.

this action would require no loss of human life. Deprived of these archetypes, their primordial divine entities could not manifest and would have no place from which to erupt. They (the indigenous peoples) would lose the matrix of their maternities and divine paternities, the sense of their origins. To them (the indigenous peoples), water underlies and precedes all creation and all forms, assuming the role of world mother, whereby the land produces live forms, personifying the figure of life's mother and mankind.[19] Without the sky, land, and water, automatically, we would eliminate the air that sustains the communication between these three agents of nature. This air constitutes the source of the word and the receptacle of translucence, brightness, the cold… and of heat, according to the cultures of these peoples.[20] It would mean the apocalypse to them, because the mythical and the profane functions of skies, waters, and lands, with all the other complements, ponderable or not, fused in these entities, are intertwined dynamically with the pleasures, desires, and the spiritual and material necessities of these peoples. Without this other half, they would not survive; they would succumb rapidly.

If this strategy is unsuccessful, we will use Plan II. We will drop a neutron bomb on the forest because its effects are swift, it does not pollute, and it eliminates life without destroying physical structures. Amazonia's biodiversity would be restored quickly by the emergent biotechnical industry and there would be 30 million inhabitants less on earth, completes the young general Greenworld, a sort of twenty-first-century Buffalo Bill. Greenworld, militant general and commander of countless environmental wars, hoped to crown his career with the invasion and ownership of the world's most impenetrable and coveted forest. Greenworld was just checking the last minute details of the invasion to activate Plan I.

* * *

After opening the General Assembly of the Peoples of Amazonia, Caloã, president of the World Confederation of Indigenous Peoples, explains that the gravity of the situation demands the presence of us all, important leaders of indigenous peoples, representatives of our peoples, of our cultures, and of our nations. We express our gratitude for the presence and solidarity of representatives from Africa, Asia, Australia, Oceania, and America who have come to our aid. A month ago, we made a last appeal to governments, churches, and multilateral organizations, which I will begin by reading to you, in particular the letter sent to the Holy Father:

19. In Amazonia, there are situations to different peoples in which these roles are reversed.
20. In this free adaptation, theoretical elements have been used taken from pages 151–153, 199–201, 261, and 281 of Gilbert Durant's book, '*Les structures anthropologiques de l'imaginaire.*' (1992). Paris: Dunod.

Great leader from the Vatican:

Today, reflecting on our shadowy past, our eyes find only death, destruction and exploitation of indigenous peoples, effected in the sophisticated form of great multinational projects, that, under the pretext of development, actually contribute to the genocide of hundreds of indigenous communities in this country and in the world. Highways, hydroelectric plants, industrial mining and military projects are built to foment invasions, diseases, hunger, poverty and death. In the last decades, the detainers of economic and political power invested in the disarticulation of indigenous peoples in order to dominate and dispossess them, as well as in the non-demarcation and non-guarantee of our lands. Alleging modernity, technology, and progress they criminally invade our territories, murder our leaders, poison our rivers, destroy our environment and treat us as inferior races, as foreigners inside our own lands ... Therefore, Holy Pontiff, we want Your Sanctity to take our clamorous shout for justice to the rest of the world, so that all may be aware that the indigenous nations are disappearing from the face of the earth.

We also hope for your support against this great crusade by governments and international armies to expel us from our houses, villages, territories and our Amazonia. And Holy Pontiff, your support is sorely needed in order to avoid more genocide against our children, our people, our cultures, and against mankind.

Our organizations confirm during the first contacts between the indigenous peoples and the colonizers, we numbered around 6,800,000 inhabitants in the Amazonian region, of which 5,000,000 occupied the Amazonian basin and 3,800,000 an area that corresponds, currently, to Brazilian Amazonia (...).[21]

Today, we are 225 indigenous peoples speaking a total of 225 different languages. These peoples live in 775 areas on Brazilian territory. Our organizations also detect the existence of 900 Indians belonging to peoples still not contacted, 192,000 living in cities and 358,310 living on their lands. Holy Father, it is important to point out that 503 indigenous lands are still not officially recognized, and a further 658 have been subject to multiple invasions. We also recorded the occurrence of 1,119 cases of indigenous killings since 1985, of which 135 cases in 2018.[22] These indigenous were killed for trying to protect the forests.

Another point to remember is that even when an indigenous group has lost its language, its religion and its traditional economy, it is still indigenous simply because it maintains its ethnic identities.[23]

21. Free adaptation of the text taken from Cimi-CNBB, 2013. In M. C. Silva (2013). 'Metamorphoses of Amazonia,' p. 101. Manaus: Editora Valer.
22. Free adaptation from Marilene Corrêa da Silva's book. Silva, M. C. (2012). '*The Country of Amazonas*," p. 126. Manaus: Editora Valer.
23. By means of an essay published on December 7, 2000, Lieggot and Lacerda make a dramatic complaint about the destruction of indigenous peoples from Amazonia during the Brazilian

Holy Father, these same invaders, after they destroyed their own natural environment, on the pretext of protecting our nature, joined forces with ecologist factions, agents of great financial groups, for the purpose of marketing the active ingredients and the secrets of trees, plants, animals, and of our own human life.

As Your Holiness knows, for us nature cannot be bought and sold, it has no economic value, it is an integral and indivisible part of our cultural identity. As Your Holiness also knows we abominate the greed, lechery and misery of these gentlemen that sow war and suffering.

These same gentlemen have conditioned mankind's survival to the internationalization and privatization of our forests and rivers, and finally, of our physical and spiritual existence. In their greed to multiply their profits, they are not interested in building a fraternal and multicultural world, free of prejudice, discrimination, social inequalities and pollution. They choose to ignore the prejudicial actions of their conceptions of the permanence of life and nature as if the ownership of a region could hide and justify them.

Holy Father, according to the Food and Agriculture Organization of the United Nations (FAO), in the year 2000, about 35,000 children died of hunger in poor countries, due to economic models that concentrate and exclude, conceived and established by these lords of war.

The disappearance of cultures that go back millennia is another perverse inheritance of this oppressive political model. The linguist Michael Krauss (2000) estimates that half of the 6,000 languages spoken at the start of twentieth century have disappeared; 80% of those that have survived are spoken only by small ageing groups.[24]

We reiterate our appeal for your intervention in the construction of this worldwide synergy on behalf of our survival as well as in the publication and search for new partners committed to peace. We have also sent a copy of this letter to all religious leaders, governments, civil organizations and information networks.

government under Fernando Henrique Cardoso (Lieggot and Lacerda, 2000). In 2019, the Indigenous Missionary Council of Brazil (CIMI) published an important report on the violence against indigenous peoples in Brazil during 2018 and 2019 (CIMI, 2019).

24. Alcida Rita Ramos (1991), one of the leading Brazilian anthropologists, has led a national and international campaign for the demarcation of indigenous lands in Amazonia. She is a defender of the physical and cultural heritage of the Yanomami indigenous people in Amazonia. This people is one of the oldest in the world. Highligt for Claudia Andujar. She is a Swiss-Brazilian artist and ativist who, for over five decades, photographed and accompanied the Yanomami and indigenous people who live in Amazonian rainforest on the border between Brazil and Venezuela, in the struggle for their rights. She was born in Switzerland in 1931, and has lived in Brazil since the 1950s. "Her artwork, The Yanomami Struggle, was exhibited at the Fondation Cartier, Paris, on January 30– May 10, 2020. It will be presented in Milan from Fall 2020 onwards. The exhibition will also travel to the Fotomuseum Winterthur (Switzerland) from June 6, 2020, and to Foundation Mapfre (Spain) from February 11, 2021" (Fondation Cartier, 2020).

Amazonas, January 1, 2019

Signed: Caloã, President of the World Confederation of Indigenous Peoples

Since one month has passed without any concrete answer, and the invaders meet in readiness for action, I, Caloã, summon and request that you as a matter of absolute urgency, discuss and formulate strategies to protect the integrity of our peoples and our territories.

* * *

After general discussion, Caloã reads to all those present the decisions of the Assembly:

In the face of the constant aggression leading to the possibility of our extinction, we have decided:

Plan I: With the help of our shamanistic spirits we will transport Amazonia and everything it contains and involves to our sacred places, for the time necessary for our self-preservation.[25] We will communicate to other peoples that during this period an average amount of water of 12×10^{12} m^3 (12,000 trillion of liters of water) will cease to circulate daily through the terrestrial atmosphere. Water enough to supply the planet's daily consumption.

We will also inform that a great amount of heat, equivalent to 400 million megawatts- due to the conversion of tropical humidity into rain- more than the electric power consumed daily by the earth's population, will cease to circulate in the terrestrial atmospheres. We believe this phenomenon will cool the relationships among the peoples and will hinder the full splendor of the germination of vegetable life in the fields. Finally, we will communicate to the great world leaders that during the period of Amazonia's absence, an amount of 0.5 gigatons (500 million tons) of carbon dioxide will cease to be absorbed annually by Amazonia's ecosystems. This will certainly accelerate the greenhouse effect, provoking incommensurable problems to the different societies worldwide.

We could produce a lengthy list of other problems that will affect mankind with the decision we are taking to protect Amazonia, but we hope these will be enough for wisdom to prevail in the minds of the invaders.

Plan II: We will take a long trip with our peoples. We will take with us the matrix of animal and vegetable life; the forest's geometric forms and its flavors,

25. Caloã also chooses the leaders who will participate in the group that will evoke the spirits that will guide the transport of Amazonia through temporary-space contractions and expansions of the Amazonian universe.

sweet, bitter, and all the other combinations; some thought patterns and imaginary colonialist typologies of White men that only we know how to decipher; the diversity of our sociologies and anthropologies, and the poetries and songs of our cultures, the trees, the birds, the fish, and the animals.

It is also our right to take all sounds and silences of the forest and the spirits of everything that exists in nature because they are creative and creations of our peoples. We will carry away the moon, the sun, and the stars of the firmament that have accompanied us during all our spiritual existence. We will also capture and keep, for whatever time necessary, the nights and all the forest's mirrors.

A long emptiness will fill the region. If the White man wants to, let him try to build his own Amazonia and illuminate it with his knowledge and his greed. We refuse to participate in this amoral, profit-driven enterprise, with roots in a morbid rationalism and autistic attitudes. Nevertheless, we communicate that we will resist the invasion by the White man, simply because Amazonia is the entrance and exit portal of the destination of our journey. A portal that constitutes a world where everything began and certainly where everything can end.

In October 2019, at the end of the work of the Synod of Bishops of Catholic Church, Holy Father created in the Vatican the Department of Amazonia dedicated to developing projects, programs, and strategies to protect it and its native peoples.

* * *

After a long absence, Amazonia bursts again onto the world scenario. Year 2050, in Manaus, capital city of the State of Amazonas and center of Amazonia's forest, the electronic screens of the great biotechnology laboratories and General Greenworld's market of ecological products are confused with the beautiful Parliament of the World Confederation of the Indigenous Peoples of Amazonia.

New alliances between indigenous peoples and capitalist economic forces are happening. Amazonia is already considered a world heritage site. Large environmental areas of Amazonia, Central Africa, and Southeast Asia are already considered world heritage sites. Their economic exploitations are fully regulated according to international standards. The new socioeconomic configuration of these regions constitutes a clean political break with current models of predatory and excluding development.

The University of Indigenous Peoples flourished and spread throughout the Amazonia, its main structural axes being: indigenous knowledge and public policies; societies of Wisdom and ethnosciences; culture, nature, and indigenous arts; indigenous economy and technological innovations; and indigenous education, science, and technology platforms. The University constitutes an important reference point for the sustainable development of the region.

Sustainability is the main utopia of the twenty-first century. It proposes a more fraternal, just, and equitable world. It apprehends the differences and cultural diversities with the prospect of building a more humane and sensitive world. It urges individuals, societies, peoples, and governments to invent new forms of coexistence and interpersonal cooperation. This book has shown the great challenges that sustainability presents to mankind in this century. It reaffirms that hope lies in education, as a civilizing agent, to improve man and societies and to solve mankind's great problems. It also reaffirms Amazonia's strategic role and sustainability for us all, from a collective and humane perspective.

References

Abrantes, P. C. C. (1998). *Imagens de Natureza, Imagens de Ciência (Images of Nature, Images of Science)*, pp. 44–52. Campinas: Editora Papirus.

Batista, E., Cavalcanti, R. B., and Fujihara, M. A. (2005). *'Caminhos da Sustentabilidade no Brasil (Paths of sustainability in Brazil)'*, pp. 122–34. São Paulo: Terra das Artes Editora.

Castel, R. (1995). *Les métamorphoses de la question sociale*, pp. 204–22. France: Gallimard.

CIMI-Conselho Indigenista Missionário do Brasil. (2019). *Relatório sobre a Violência contra os Povos Indígenas no Brasil, Dados de 2018 (Report on Violence against indigenous peoples in Brazil – Data 2018)*. Brasília: CIMI.

Dubois, J. L., and Mahieu, F. R. (2002). 'La dimension sociale du développement durable: réduction de la pauvreté ou durabilité sociale?'. In M. Jean-Yves (ed.), *Developpement durable? Doctrines, Pratiques, Évaluations*, pp. 73–94. Paris: IRD Éditions.

Durant, G. (1992). *Les structures anthropologiques de l'imaginaire*. Paris: Dunod.

Durkheim, E. (2003). *The elementary forms of religious life*. Oxford: Oxford University Press.

Elias, N. (1994). *O Processo Civilizador. Formação do Estado e Civilização (The civilizing process. Formation of the state and civilization)*, pp. 23–85. Rio de Janeiro: Jorge Zahar Editor.

Fondation Cartier. (2020). 'Claudia Andujar – Exposition 30 Janvier – 10 Mai 2020 – La Lutte Yanomami.' Exposition organisée dans le cadre du partenariat entre la Fondation Cartier et Triennale Milano. Paris: Fondation Cartier pour l'art contemporain.

Freitas, M., and Silva Freitas, M. C. (2014). 'Fragments of utopias from twenty-first century and Amazonia: projections and controversies', *Innovation: The European Journal of Social Science Research*, 27 (3), 275–94.

Freitas, M., and Silva Freitas, M. C. (2016). *'A sustentabilidade como paradigma: cultura, ciência e cidadania (Sustainability as paradigm: culture, science and citizenship)'*. Petrópolis: Editora Vozes.

Freitas, M., Silva Freitas, M. C., and Marmoz, L. (2003). *A ilusão da sustentabilidade (The illusion of sustainability)*. Manaus: Editora Valer.

Freitas, M., Silva Freitas, M. C., and Meunier, O. (2016). *Amazônia: passado-presente-futuro (Amazonia: past-present-future)*. Curitiba: Editora Juruá.

Freitas, M., Silva Freitas, M. C., Ioris, A., and Castro Júnior, W. E. (2017). *Amazônia (Amazonia)*. Lisboa: Chiado Editora.

Green Left–wekly. (1993) 'Vietnam: war and the environment', *Greenleft*, 13 July, <https://www.greenleft.org.au/content/vietnam-war-and-environment>, accessed 12 July 2019.

Haberlandt, M. (1929). *Etnografia, Estudio General de las Razas (Ethnography, general survey of races)*, 2nd edn., pp. 29–30. Barcelona: Editorial Labor.

Holden, E. (2019). 'Trump begins year-long process to formally exit Paris climate agrément', *The Guardian*, 5 November, p. 1, <https://www.theguardian.com/us-news/2019/nov /04 / donald-trump-climate - crisis-exit-paris-agreement>, accessed 5 November 2019.

Homma, A. K. O. (2003). História da Agricultura na Amazônia: da era pré-colombiana ao terceiro milênio *(History of agriculture in Amazonia: from the pre-Columbian Era to the third Millennium)*. Brasília: Embrapa Informação Tecnológica.

Ianni, O. (1999). *Metáforas da Globalização—Teorias da Globalização (Metaphors of Globalization—Theories of Globalization)*, pp. 24–8. Manaus: Editora da Universidade Federal do Amazonas, Manaus.

Krauss, M. (2000). Citation. In P. A. Cox (2001). 'Essay on science and sociology: Tribal will knowledge survive the millenium?' *Science Magazine*, 287 (5450), 44–5.

Larrère, C., and Larrère, R. (1997). *Du bom usage da la nature*, pp. 292–93. Paris: Alto Aubier.

Lecteur de Pic, C. A. (2005). 'Secrets de la nature et mystères divins'. In D. Courcelles (ed.), *D'un principe philosophique à un genre littéraire: les secrets*. Paris: Honoré Champion Éditor.

Lieggot, R., and Lacerda, R. (2000). 'Governo FHC: mais dois anos de engodo. Em defesa da causa indígena (FHC Government: two more years of deception. In defense of the indigenous cause)', *Jornal Porantim, Brasília*, 7 December, pp. 8–9.

Mantoux, P. (1927). *A Revolução Industrial no Século XVIII (The industrial revolution in eighteenth century)*, pp. 177–315. São Paulo: Editora Hucitec.

Marx, K. (1984). *O Capital (The capital), vol.* I, *tomo 2*, p. 17. São Paulo: Editora Abril Cultural.

Marx, K. (1985). *O Capital – Crítica da Economia Política (The capital—critique of political economy), vol.* III, *tomo 2*, p. 237. São Paulo: Editora Abril Cultural.

Ramos, A. R. (1991). 'Por que demarcar as terras Yanomami? (Why demarcate the Yanomami' lands?)', *Informativo da Funai e Cimi*, Brasília, pp. 1–4.

Reid, M. (1995). 'O impacto da conservação da biodiversidade sobre os povos indígenas (The impact of biodiversity conservation on indigenous peoples)'. In A. Lopes, and L. Donisete (eds.), *A temática indígena na escola (The indigenous thematic in school)*, p. 109. Brasília: MEC/MARI, UNESCO.

Renan, E. (2005). 'As representações científicas da Amazônia: o lugar das etnociências (The scientific representations of Amazonia: the place of ethnosciences)'. In M. Freitas (ed.), *Amazônia: a natureza dos problemas e os problemas da natureza (Amazonia: the nature of the problems and the problems of nature)*, vol. I, pp. 181–196. Manaus: Editora da Universidade Federal do Amazonas.

Rudé, G. (1995). *Ideology and popular protest*. USA: University of North Carolina Press.

Schellnhuber, H. J. S., Messner, D., Leggewie, C., Leinfelder, R., Nakicenovic, N., Stefan, R., Schlacke, S., Schmid, J., and Schubert, R. (2011). *World in transition – A social contract for sustainability*. Berlin: German Advisory Council on Global Change.

Shiraishi Neto, J. (2007). *Direito dos Povos e das Comunidades Tradicionais no Brasil (Right of peoples and traditional communities in Brazil)*, pp. 201–11. Manaus: Editora da Universidade do Estado do Amazonas.

Silva, M. C. (2012). *O Paiz do Amazonas (The country of Amazonia)*. Manaus: Editora Valer.

Silva, M. C. (2013). *Metamorfoses da Amazônia (Metamorphoses of Amazonia)*. Manaus: Editora Valer.

Spurgeon, D. (2003). 'Traces of agent orange are found in food in Vietnam-NCBI', *BMJ* 327 (7414), 521, <https://www.ncbi.nlm.nih.gov/pmc/articles/PMC193378>, accessed 13 July 2019.

UNESCO. (2005). *'Rapport Mondial de L'UNESCO, Vers les Sociétés du Savoir (UNESCO's world report, towards knowledge societies)'*, pp. 190–1. Paris: Éditions UNESCO.

Vivien, F.-D. (2001). 'Histoire d'un mot, histoire d'une idée: le développement durable à l'épreuve du temps'. In M. Jollivet (ed.), *Le développement durable, de l'utopie au Concept*, pp. 19–60. Paris: Elsevier.

Wagner, A. (2019). *Antropologia da Amazônia (Antropology of Amazonia)*, pp. 64–7. Manaus: UEA Edições.

Wahl, P. (2001). 'Gouvernance mondiale (Global governance)'. In B. François (ed.), *Dictionnaire critique de la mondialisation (Critical dictionary of globalization)'*, pp. 173–6. Paris: Le Pré aux Clercs.

Yale E360. (2010) 'In war-scarred landscape, Vietnam replants its forests', *Yale E360*, 4 November, <https://e360.yale.edu/features/in_war-scarred_landscape_ vietnam_ replants_ its _ forests>, accessed 10 August 2019.

CHAPTER NINE

Preservation of Amazonia: Commitments for Us All

9.1 Introduction

The ongoing pillage of regional and planetary environments has stimulated the creation of environmentalist movements and political actions by organized groups (Robbins, 2012). Absence of permanent public policies in poor regions and countries leads to speculations and challenges for sustainability. In the technological field, in order for sustainability to be technically operational, absolute priority must be given to replacing the current energy matrix. It also imposes a fundamental change in educational policies, at all hierarchical levels, and requires the formulation of novel paradigms for public policies and the sciences, with impacts on the processes of organization and application.

Planet's ecological destruction puts new challenges for mankind. Amazonia is inserted at this conjuncture as an important reference for mankind and the planet's sustainability.

9.2 Sustainability and the Preservation of Amazonia

There is no vaccine to immunize the planet of man's deleterious actions and the destructive productive forces. The planet's sustainability is a collective and shared historical construction.

Sharing sustainability through public policies also has impacts on the privatist and alienating civilizing concept. Society and the market have encouraged speculation on the senses and operationalities of this notion. Twenty-first century's concern with sustainable development raises six main issues, all related to the accumulation, expansion, and circulation of capital (Freitas and Silva Freitas, 2013a).

The first is symbolic, and therefore the most complex. There is a certain illusion about the notion of sustainability, which considers that the mechanisms that make it operational do not establish "where, when, and how" to replace the classical form of development. It is possible that the wait will be for something that will never happen; we run the risk of constructing a socioeconomic enterprise that is so unsound that it will never materialize.

Second, there is an incompatibility between the notion of sustainability and the concept of growth- not financial growth, rather the growth of the flow of mass and energy. This will result in prioritizing the marketing of goods with greater durability and changing the world's industrial matrix, implementing clean development mechanisms.

The third problem concerns the dynamics surrounding the process of financial accumulation. Core countries are becoming richer and the peripheral countries are becoming poorer. This situation demands the incorporation of the notion of sustainability, including those essential requirements for a basic standard of living. There is also an additional problem: the increasing wave of privatization of production conspires against the management of the planet's natural wealth.

Fourth, hypocrisy exists within central governments. History has recorded the discourse of governments that oppose practical action. They will not act in any way that might risk the welfare of their voters and their economic and political stability.

Fifth, the notion of sustainable development has historical validity only in local experiments. Yet, there is a common objective: the preservation of biodiversity associated with cultural diversity. The objective conditions for preservation remain controversial. Disregard for the biosphere and the standard capitalist approach to exploiting natural resources obstruct attempts to find solutions.

Finally, there is an increasing tension between the notion of sustainability and the universal principle of national security. The friction generated will depend on the evolution of political processes on a worldwide scale.

These six issues which underlie worldwide sustainability processes are present in Pan-American Amazonia. They are directly related to the new forms of concentration, organization, reproduction, and circulation of capital led by transnational groups and central countries. The contributions of education, science, and technology should accelerate and qualify this new "worldwide enterprise" toward sustainable processes (Silva Freitas et al., 2015). They will also contribute to incorporating Amazonia into the future of mankind, which has a growing tendency to be technified and ecologized.

Amazonia's future is hostage to three major tendencies: the first projects its increasing ecological destruction by agribusiness, cattle ranching, large-scale logging, and massive fires. The second predicts an increasing number of sustainable development programs and projects in the region, reaffirming its ecological and symbolic importance to Brazil and to mankind. Finally, the third tendency presupposes greater international control over its preservation and a spreading world collaboration in its development through international economic and political protocols and agreements. Scenarios combining elements of these three tendencies, in contextualized form, are also feasible.

The development of Amazonia is a challenge for us all. It is very dependent on technological innovations and the Brazilian government's political decisions. To institutionalize public policies fully in Amazonia requires implementing sustainable development models integrated with its cultural, ecological, and socioeconomic complexities. It also requires regional and national integration and the organization of social structures and technologies accessible to all, generating income, social inclusion, and citizenship for its populations, as well as environmental preservation.

For these reasons, the following political commitments to Amazonia are imperative: mobilization of Brazilian society with a view to reaffirming the importance of education, science, and technology as processes of humanization and socioeconomic development of Amazonia and Brazil; investment of US$1 trillion in science, technology, and innovation policies, as a contribution to the regional and national integration of Amazonia to Brazil`s national Project, during 2020–2030; guarantee of sovereignty and institutionalization of the state's presence in the region, through integration, decentralization, and internalization of state and federal agencies to plan and execute public policies (Silva Freitas et al., 2015); implementation not only of centers for the diagnosis and control of illegal deforestation and land use, but also of public policies for the integration of environmental services to all Amazonia, through the recovery of degraded areas, biodiversity conservation, water resources, and mitigation of climate change; guarantee of scientific education and constitutional rights for indigenous peoples and traditional communities, and promotion of social equity, considering gender, generation, race, social class, and ethnicity; and the strengthening of cooperation between Brazil and the other Amazonian countries through science and technology programs.

Homma (2003) presents strategic themes for the region's sustainable future. He analyzes the historicity of economic cycles in Amazonia and its world nexus. His approach prioritizes facts, characters, scientific achievements, technological innovations, movement of genetic resources, institutions, and public policies that have marked the use of land and the history of agriculture in the Amazonian region. He proposes feasible solutions for occupation, use, and the sustainable development of the region, contextualized and integrated with the interests of its populations. Brazil's future is very dependent on Amazonia's sustainable development. This type of development in the region depends on the impact of education, science, and technology on its public policies. The ecological destruction of Amazonia ends all possibility of Brazil becoming a developed country, capable of assuming new challenges and international commitments.

Brazilian governments and productive forces have no idea on how to develop Amazonia. It is still being held hostage by a backward and uncompromising elite, along with Brazil and mankind. Amazonia has no shortage problem; on the contrary, its main feature is its cultural and environmental immensity. At this point in time, Brazilian Amazonia's future is a worldwide tragedy foretold. The global alignment of political, economic, and religious forces dedicated to the protection of Amazonia is ongoing.

Political actions to designate Amazonia as a world heritage site, to guarantee its protection by international forces, and to implement its sustainable use start to gain predictable contours.

References

Freitas, M., and Silva Freitas, M. C. (2013a). 'Sustainability, Amazonia and environment: propositions and challenges', *International Journal of Environmental Studies*, 70 (4), 467–76.

Freitas, M., and Silva Freitas, M. C. (2013b). *Sustainability; Man-Amazonia-World.* Baltimore: America Star Books.

Homma, A. K. O. (2003). *História da Agricultura na Amazônia: da era pré-colombiana ao terceiro milênio (History of agriculture in Amazonia: from the pre-Columbian Era to the third Millennium).* Brasília: Embrapa Informação Tecnológica,

Robbins, P. (2012). *Political Ecology-A Critical Introduction*, 2nd edn., p. 58. Chichester: John Wiley & Sons Ltd..

Silva Freitas, M. C., Freitas, M., and Ruiz, M. A. (2015). 'La politique publique pour le développement durable: l'Amazonie, un cas illustrative', *Innovation: The European Journal of Social Science Research*, 28 (2), 192–210.

PETER LANG
PROMPT